Fluid behaviour in biological systems

LEONARD LEYTON

FELLOW OF LINACRE COLLEGE, UNIVERSITY OF OXFORD

Fluid behaviour in biological systems

CLARENDON PRESS. OXFORD
1975

Oxford University Press, Ely House, London W. 1

GLASGOW   NEW YORK   TORONTO   MELBOURNE   WELLINGTON
CAPE TOWN   IBADAN   NAIROBI   DAR ES SALAAM   LUSAKA   ADDIS ABABA
DELHI   BOMBAY   CALCUTTA   MADRAS   KARACHI   LAHORE   DACCA
KUALA LUMPUR   SINGAPORE   HONG KONG   TOKYO

ISBN 0 19 854126 0

QP
90.5
L49

PRINTED BY
THE UNIVERSITIES PRESS
ALANBROOKE ROAD, BELFAST BT69HF, NORTHERN IRELAND

Preface

THE idea of writing this book arose mainly from courses of lectures on Physics for Biologists and on Environmental Physics that I have given for several years to first-year biology undergraduates. Similar courses now given in most universities recognize the importance of a knowledge of physics in modern biology, but in my experience, students do not always take very readily to the subject. This may be due in part to the difficulties that many biologists experience with physics and especially with the associated mathematics, but I believe that important contributory factors are teaching methods and, in particular, the lack of suitable texts.

There are, of course, a number of publications available nowadays which give accounts of general physical principles written especially for biologists, and there are also numerous, more specialized, texts in which these principles are applied to particular fields. In the case of many important topics however, there appears to be little written at an intermediate level. We might take as an example the treatment of boundary layers which play a key role in several aspects of the relationship between an organism and its environment. In the more general texts, these are usually dismissed in a paragraph or so, whilst the more advanced texts generally assume a familiarity with the basic principles that few biologists possess. In fact, most of the basic principles of boundary layers are only adequately explained in engineering texts on fluid mechanics, few if any of which are to be found in biological libraries.

One of the main aims of this book has been to provide a reasonably comprehensive account of the principles of fluid mechanics and their biological applications. Since all organisms live in a fluid environment and several important physiological processes involve fluid behaviour, an understanding of these principles would appear to be essential for all biologists.

I have interpreted 'fluid behaviour' and 'biological systems' in their widest context so that besides the transport of heat, liquids, and gases within organisms and between organisms and their environment, I have added a chapter on locomotion including some elementary hydro- and aero-dynamics. Also, bearing in mind the difficulties biologists have in grasping the principles of environmental physics, it appeared logical to extend the account to the environment and to introduce some elementary micrometeorology and microclimatology. In order to make this reasonably complete I have included radiation and heat balances which, strictly speaking, are outside the context of my title.

One of the most difficult problems was to decide upon a degree of detail, which, whilst still making intelligible reading for undergraduates, would serve also as a source of reference and, for more senior biologists, as an introduction to the more advanced literature. With the problems of the latter in mind I have included rather more detail than may be thought necessary for undergraduates. For the benefit of animal physiologists and medical students I have given a fairly lengthy account of blood flow and for physiologists in general, an introduction to the thermodynamics of water transport. I would emphasize that in no case have I assumed a greater physical or physico-chemical knowledge than most biologists usually possess.

The problem of how much mathematics to include was even more difficult. This is essentially a quantitative subject in which relationships are most effectively expressed in mathematical terms, and I offer no apologies for what may appear to be a rather high content of mathematical formulae. However, no more than an elementary knowledge of calculus is required of the reader and where appropriate, I have explained the relationships in words. To reduce still further the mathematical content of the text, I have relegated most of the derivations (plus a few other items) to appendices. I would strongly advise readers to study these derivations, not only to improve their knowledge of the underlying principles, but also to appreciate the various assumptions that may have been made in obtaining a particular solution. Far too many unsuspecting biologists in the past have failed to understand the limitations imposed by these assumptions when applying mathematical models to biological problems.

In the examples I have selected to illustrate the various physical principles, I have tried to strike a fair balance between the plant and animal kingdoms. Naturally, some of these examples, such as phloem and xylem transport and much of the discussion on water relations and environmental physics will be of greater interest to plant physiologists and ecologists than to their zoological colleagues. Conversely, the chapter on locomotion and the account of blood flow may only prove of interest to zoologists. However, I would emphasise that most of the physical principles I have attempted to explain are of general biological application and I hope that their relevance will be appreciated by all biologists irrespective of specialization.

Oxford, 1974   L. LEYTON

Acknowledgements

I wish to express my sincere thanks to those of my colleagues who willingly gave of their time in advising on the subject matter and in reading and criticising the drafts, in particular to Peter Nye, Evan Reynolds, Frank Thompson and Colin Wood of Oxford University and to Margot Roach of the University of Western Ontario.

Thanks are also due to the following for permission to reproduce the material indicated or to use it in preparing illustrative material:

I. M. Klotz (Fig. 1.2), W. Drost-Hansen (Fig. 1.3), J. Cronshaw (Fig. 2.2), M. J. Canny (Fig. 2.3), J. A. Petty (inset, Fig. 3.3), B. A. Meylan (Figs. 3.3, 3.4), W. Nachtigall (Figs. 5.7, 6.10), R. H. J. Brown (Fig. 6.11), C. J. Wood (Fig. 6.17), R. Yakuwa (Fig. 7.4), R. T. Tregear (Fig. 7.6), J. L. Monteith (Figs. 7.10, 9.6), The Director, NERC Institute of Hydrology (Fig. 9.9), G. Szeicz (Fig. 9.12), The Director, Swiss Federal Forest Research Station (Figs. 9.14, 9.15), A. C. Burton (Fig. 10.5); John Wiley and Sons (Figs. 1.2, 7.7), American Chemical Society (Fig. 1.3), Academic Press (Fig. 2.2), American Physiological Society (Figs. 2.5, 7.6), The Royal Society (inset, Fig. 3.3), Royal Microscopical Society (Fig. 3.4), Springer-Verlag (Figs. 5.7, 6.10), Weidenfeld Publishers (Figs. 6.4, 6.6), Cambridge University Press (Fig. 6.11), Company of Biologists (Figs. 6.13, 7.9), Japanese Meteorological Agency (Fig. 7.4), Edward Arnold (Figs. 7.10, 9.6), University of Chicago Press (Fig. 9.1), American Geophysical Union (Fig. 9.12), Clarendon Press (Fig. 9.13) and Year Book Medical Publications (Fig. 10.5).

Contents

FIGS. 2.2, 2.8, 3.3, 3.4, 3.5, 5.7, 8.1, and 9.9 appear as a separate plate section between pp. 26–7

A note on dimensions, units, and symbols

Dimensions

EXCEPT in a few special cases, all the quantities mentioned in this text may be expressed in terms of one or more of the following five basic dimensions: length (L), mass (M), time (T), temperature (θ), and electric current (I). The dimensions of some of the more commonly used quantities involving M, L, and T are given below to illustrate their derivation:

area ($L \times L = L^2$); volume ($L \times L \times L = L^3$);
velocity ($= \text{distance}/\text{time} = L/T = LT^{-1}$);
acceleration ($= \text{rate of change of velocity with time} = LT^{-1}/T = LT^{-2}$);
force ($= \text{mass} \times \text{acceleration} = MLT^{-2}$);
pressure ($= \text{force per unit area} = MLT^{-2}/L^2 = ML^{-1}T^{-2}$);
energy ($= \text{force} \times \text{distance} = MLT^{-2} \times L = ML^2T^{-2}$);
power ($= \text{rate of energy production} = \text{energy}/\text{time} = ML^2T^{-3}$).

The practical importance of dimensions rests on the rule that the dimensions of the quantities on both sides of an equation must balance. This provides an invaluable check on the validity of an equation and helps in the derivation of the dimensions and therefore units, of unknown quantities. In the method of dimensional analysis, this rule is applied to the solution of problems in which the relevant parameters can be set down theoretically and then combined in such a way as to maintain the appropriate dimensional equivalence; we shall meet a number of examples of this method in the text.

Units

Following current practice, the units used here are those recommended in the Système International d'Unités (S.I.). The five base quantities corresponding to the above dimensions, with their units and symbols are:

TABLE 1.

Quantity	S.I. unit	Symbol
Length	metre	m
Mass	kilogram	kg
Time	second	s
Electric current	ampere	A
Thermodynamic temperature	kelvin	K

The thermodynamic temperature is perhaps better known as the absolute temperature.

The various supplementary and derived units which make up for the remaining quantities mentioned in this text will be found below in the table of symbols.

Although the general adoption of the S.I. system has removed much of the confusion so evident in the earlier literature, we are still faced with the problem of dealing with data and formulae inherited mainly from the centimetre–gram–second (c.g.s.) system which are still very much in common use, especially in biology. Where it has appeared desirable to maintain continuity, both S.I. and earlier units have been included.

Particular problems have arisen from the engineering literature which has provided most of the data on fluid mechanics. Though most engineers have now turned to the S.I. system, many of the standard texts (especially American ones) still maintain the custom of treating force (F) as a basic dimension. If the unit of this force is the *kilogram-weight*, i.e. kg mass $\times g$ (the acceleration due to gravity) $= \text{kg} \times 9.81$, its magnitude is 9.81 times the newton (N) which is the derived S.I. unit of force. This is because the newton is defined as the force which gives a mass of 1 kg an acceleration of 1 m s^{-2} (not 9.81 m s^{-2}). Great care must therefore be exercised when converting from engineering units. The situation is made even more difficult for the biologist when, as is sometimes the case, forces are simply referred to as kilograms, pressures as kilograms per square metre and densities include the weight rather than the mass.

Symbols

The literature still reveals a considerable variation in the use of symbols, not only between the different subjects covered in this text, but also within the same subject. As far as possible, the symbols used here correspond to those recommended either by the Symbols Committee of the Royal Society (1971) or by the British Standards Institution (BS 1991). However, where a particular branch of the subject has evolved a system of symbols with which the reader must be familiar if he is to read further, attempts have been made to include this along with the more conventional system.

Symbols and Units

A	area (m^2)
Å	angstrom (10^{-10} m)
C	mass flux (kg m^{-2} s^{-1})
C_D	average (total) drag coefficient
C_f	pipe or local friction coefficient
C_F	average (total) friction coefficient
C_L	lift coefficient
D	diffusion coefficient (m^2 s^{-1})
D	rate or shear (s^{-1})
D'	invasion coefficient (m^2 s^{-1} bar^{-1})
E	energy, work (J = N m)
E	evaporation flux (kg m^{-2} s^{-1})
E	irradiance (W m^{-2})
E	electromotive force (V)
F	force (N)
G	heat flux (storage) to ground (W m^{-2})
G	Gibbs' free energy (J)
H	heat flux to air (W m^{-2})
I	thermal resistance (J^{-1} m^2 s °C)

I	radiation intensity (J m^{-2})	Q	quantity
I	electric current (A)	R	gas constant (J mol^{-1} K^{-1})
J	Joule (kg m^2 s^{-2} = N m)	R	electrical resistance (ohm)
J	Onsager flux (various)	R_N	net radiation (W m^{-2})
K	hydraulic conductivity (m s^{-1} etc.)	S	short wave radiation (W m^{-2})
L	latent heat (J kg^{-1})	T	absolute temperature (K)
L	long-wave radiation (W m^{-2})	T	tortuosity factor
L	Onsager coefficient (various)	T	torque, moment of force (N m)
M	molecular weight (kg)	U	free stream velocity (m s^{-1})
M	metabolic heat rate (W)	V	electric potential difference (volt)
N	number	V	volume (m^3)
N	newton (kg m s^{-2})	V_M	molar volume (m^3 mol^{-1})
P	power (W)	W	weight (mg)
Pa	pascal (= 1 N m^{-2})	W	watt (J s^{-1})

a	light absorption coefficient	k	specific conductance (Onsager) (mho)
b	width, wing span (m)	l	length (m)
c	velocity of light (m s^{-1})	m	mass (kg)
c	mass concentration (kg m^{-3})	m	hydraulic radius (m)
c_M	molar concentration (mol m^{-3})	mb	millibar (1 mb = 10^2 N m^{-2})
c	Kozeny-Carman coefficient	p	momentum (kg m s^{-1})
c_p	specific heat at constant pressure (J kg^{-1} K^{-1})	p	pressure (N m^{-2} = P_a)
d	diameter (m)	q	rate
d	zero plane displacement (m)	q	specific humidity
e	vapour pressure (e.g. mb)	r	radius (m)
f	frequency (cycles s^{-1}, Hertz)	r	diffusion resistance (s m^{-1})
g	acceleration due to gravity (m s^{-2})	s	internal surface factor (m^{-1})
h	height (m)	s_0	specific surface (m^{-1})
h	relative humidity	s	distance along path (m)
h	convection coefficient (J m^{-2} s^{-1} °C^{-1})	t	time (s)
		u	velocity (m s^{-1})
$\bar{h}$	average convection coefficient (J m^{-2} s^{-1} °C^{-1})	u_*	friction velocity (m s^{-1})
		x	mole fraction
k	specific permeability (m^2)	y	activity coefficient
		z_0	roughness length (m)

α	absorptance	ν	kinematic viscosity (m^2 s^{-1})
α	thermometric diffusivity (m^2 s^{-1})	π	osmotic pressure (N m^{-2})
α	Bunsen coefficient	ρ	density (kg m^{-3})
β	coefficient of expansion	ρ	reflectance
β	Bowen ratio (H/LE)	σ	Stefan–Boltzmann constant (W m^{-2} K^{-4})
γ	psychrometer constant (e.g. mb °C^{-1})	σ	surface tension (N m^{-1})
Δ	difference, slope	σ	xylem conductivity coefficient (m^4 N^{-1} s^{-1})
δ	thickness (m)		
ϵ	porosity	σ	membrane reflection coefficient
ϵ	emmisivity	τ	shear stress, momentum flux (N m^{-2} = kg m^{-1} s^{-2})
η	dynamic viscosity (kg m^{-1} s^{-1} = Pa s)		
θ	common temperature (°C)	τ	transmittance
κ	turbulent transfer coefficient (m^2 s^{-1})	ϕ	flux (flow rate per unit area)
		ϕ	volume fraction
λ	thermal conductivity (W m^{-1} °C^{-1})	χ	mixing ratio
λ	activity	ψ	water potential (N m^{-2}, bar)
λ	mean free path (m)	ω	angular velocity (rad s^{-1})
λ	wavelength (m)	ω	mass rate of flow (kg s^{-1})
μ	chemical potential (J mol^{-1})		

S. I. Decimal multiples

Multiple	Prefix	Symbol	Multiple	Prefix	Symbol
10^{-1}	deci	d	10	deca	da
10^{-2}	centi	c	10^2	hecto	h
10^{-3}	milli	m	10^3	kilo	k
10^{-6}	micro	μ	10^6	mega	M
10^{-9}	nano	n	10^9	giga	G
10^{-12}	pico	p	10^{12}	tera	T
10^{-15}	femto	f			

Physical constants

σ	Stefan–Boltzmann	$5 \cdot 670 \times 10^{-8}$ W m^{-2} K^{-4}
g	acceleration due to gravity	$9 \cdot 807$ m s^{-2}
R	gas constant	$8 \cdot 314$ J mol^{-1} K^{-1}
atm	atmospheric pressure	$1 \cdot 013 \times 10^5$ N m^{-2} ($= 1 \cdot 013$ bar)
c	speed of light	$3 \cdot 00 \times 10^8$ m s^{-1}

Conversion factors

Length	inch $= 0 \cdot 0254$ m		*Mass*	1 pound (mass) $= 0 \cdot 4536$ kg
	foot $= 0 \cdot 3048$ m		*Time*	1 year $= 3 \cdot 156 \times 10^7$ s
Area	sq. inch $= 6 \cdot 452 \times 10^{-4}$ m^2			1 day $= 8 \cdot 64 \times 10^4$ s
	sq. foot $= 9 \cdot 29 \times 10^{-2}$ m^2		*Force*	1 dyne $= 10^{-5}$ N
Volume	cub. inch $= 1 \cdot 639 \times 10^{-5}$ m^3			1 kgF $= 9 \cdot 81$ N
	cub. foot $= 2 \cdot 832 \times 10^{-2}$ m^3		*Pressure*	1 dyne cm^{-2} $= 0 \cdot 1$ N m^{-2}
	1 litre $= 10^{-3}$ m^3 $=$ dm^3			1 ins water (4°C) $= 249 \cdot 1$ N m^{-2}
Velocity	1 ft sec^{-1} $= 0 \cdot 3048$ m s^{-1}			1 mm Hg (0°C) $= 133 \cdot 3$ N m^{-2}
	1 mile h^{-1} $= 0 \cdot 447$ m s^{-1}			1 bar $= 10^5$ N m^{-2}
	1 knot $= 0 \cdot 514$ m s^{-1}			
Energy	1 BTU $= 1055$ J		*Power*	1 BTU h^{-1} $= 0 \cdot 293$ W
	1 erg $= 10^{-7}$ J			1 HP $= 745 \cdot 7$ W
	1 calorie $= 4 \cdot 186$ J			1 calorie s^{-1} $= 4 \cdot 186$ W
	1 kWh $= 3 \cdot 6 \times 10^6$ J		*Viscosity*	1 poise $= 0 \cdot 1$ kg m^{-1} s^{-1}
	1 electron volt $= 1 \cdot 602 \times 10^{-19}$ J			1 stokes $= 10^{-4}$ m^2 s^{-1}

Mathematical symbols

$=$	equals	$<$	smaller than
$\neq$	not equal to	$>$	larger than
$\approx$	approximately equal to	$\ll$	much smaller than
$\rightarrow$	approaches	$\gg$	much larger than
$\simeq$	asymptotically equal to	$f(x)$	function of x
$\propto$	proportional to	$\exp x$, e^x	exponential of x
∞	infinity	e	base of natural logs
Σ	sum of	ln	natural log of x
		$\sqrt{(x)}$	square root of x

1. Introduction

The biological importance of fluid behaviour

ALL organisms live in a fluid environment so that any process which involves an interaction between the organism and its environment, such as movement, heat exchange or gaseous exchange, will be influenced by the behaviour of the fluid. Furthermore, several physiological processes occur in organisms that involve fluid movement, e.g. the circulation of blood and lymph, the movement of air in respiratory systems, the transport of water and foodstuffs in plants. It follows that a knowledge of fluid behaviour and of the factors that influence it, is fundamental to our understanding of these processes.

Solids and fluids

If we are to discuss the behaviour of fluids, it would appear appropriate in the first place to define what is meant by a fluid.

Fluids are distinguished physically from solids by the relative mobility of their constituent molecules. Although the molecules of a solid may vibrate or rotate, they do so about a fixed average position, whereas those of a fluid can move from one position to another within the body of the fluid. An important property of fluids which underlies much of the behaviour we shall discuss is their response to forces tending to produce a deformation, or change of shape. With typical solids, the amount of deformation is limited and as long as the force is not too great, the solid will regain its shape when the force is removed. With fluids on the other hand, deformation is continuous whilst the force is acting, and remains even when the force is removed. In other words, solids possess elastic properties; fluids do not. There are, however, many substances that appear to have intermediate properties. Some apparently solid materials never regain their shape, whilst certain fluids, including many biologically important colloidal systems, produce only a limited deformation when subjected to a small force, but suffer permanent deformation and flow when the force is increased. We shall discuss these apparently anomalous substances in Chapter 10; for the present we shall concern ourselves with the more typical fluids.

Fluids comprise both liquids and gases. In liquids the energy of the molecules associated with their movement is generally less than the energy attracting them to each other so that the average distance between molecules is relatively small. In gases the molecules are more widely distributed, the attractive energy between them is less than their energy of movement, and therefore they can move more freely than in a liquid. Gases and liquids also differ in their compressibility. Whereas the volume of a gas changes with pressure (Boyle's law) that of a liquid does not, at least over the normal range of pressures. In most of the following calculations however, it has been assumed for the sake of simplicity, that even gases are incompressible; this introduces relatively little error because, for the kind of behaviour concerned, compressibility only becomes important at very high velocities, of the order of one-fifth the velocity of sound.

The structure of water

The behaviour of a fluid can usually be related to its molecular structure, but the structure of liquids in general is less well understood than that of gases; water presents a particular problem since it possesses many unique properties that distinguish it from other liquids. These arise from the unusual structure of the molecule in which the O and H atoms are arranged at the apices of a triangle (Fig. 1.1a). When two H atoms combine with an O atom to form water, they do so by covalent bonds; each H electron pairs with an electron in the oxygen's outer shell of 6 electrons, leaving two pairs of electrons in the latter (so-called lone pairs) with orbits extending tetrahedrally (Fig. 1.1b). The result is a tetrahedron with two positively charged sites and two negatively charged sites at the corners. The water molecule therefore possesses electrical polarity (hence the dipole structure) and will tend to attach itself by H-bonds to four neighbouring molecules, two at the positively-charged sites and two at the negatively-charged sites.

The H-bonds provide a fairly strong attachment between water molecules and give rise to the various crystalline structures of ice. There is, however,

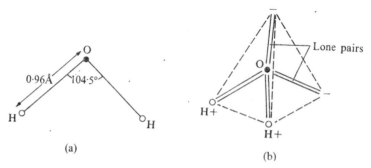

Fig. 1.1. Diagram of water molecule showing (a) bond angle and length, (b) tetrahedral arrangement of electrical charges.

no generally accepted model for the structure of liquid water. It is believed to possess a quasi-crystalline structure with extensive H-bonding between the molecules, but of a very impermanent nature. Broadly speaking, the proposed models fall into two groups, one envisaging discrete but short-lived molecular clusters (e.g. the flickering cluster concept), the other a random network of closed molecular rings or cage-like (clathrate) structures.

The polar nature of the water molecule renders it very susceptible to electrical charges. For example, in the presence of positively charged ions, several molecules (the number depending on ion species) become firmly

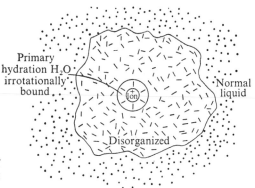

Primary hydration H₂O irrotationally bound

Normal liquid

Disorganized

FIG. 1.2. Model of the structure of water in the neighbourhood of an ion. After Klotz (1970).

attached by H-bonds to form a *hydration shell* in which they take up a distinct orientation (Fig. 1.2). An important feature of many of the systems with which we shall be dealing is the occurrence of a similarly orientated layer of water molecules (so-called *ordered water*) on hydrophilic (i.e. electrically charged) surfaces, again by H-bonding. This applies particularly to the surfaces of colloidal particles such as protein, cellulose, and other macromolecules (also clays) and accounts for several of the properties characteristic of colloidal suspensions (Bernal 1965). Although there appears to be little evidence for an ice-like structure as certain authors suggest, there is little doubt that the properties of this ordered water are very different from those of the bulk water further away from the surface e.g. it has a much higher viscosity and a lower solubility and hence diffusivity for ions.

To what extent the surface effect persists depends on the nature of the surface. According to Drost-Hansen (1969) the situation for hydrophilic surfaces is somewhat similar to that illustrated for ions in Fig. 1.2, namely a few layers of highly ordered molecules immediately next to the surface (of the order of 10 to 20 Å or 1 to 2 nm thick), an intermediate zone in which the structured arrangement decreases and local disorder increases, finally merging into the bulk water further away (Fig. 1.3).

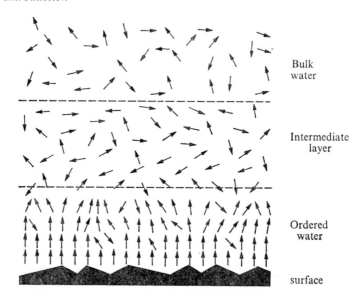

FIG. 1.3. Hypothetical orientation of water molecules near a polar surface. Note the difference between the ordered layer next to the surface and the bulk water further away. Redrawn with permission from Drost-Hansen (1969). *Ind. Engn. Chem.* **6,** 10. Copyright by the American Chemical Society.

Boundary layers and viscosity

Let us now turn to some of the phenomena associated with the flow of fluids. If we were to measure the speed of say, water flowing over a smooth flat plate, we would find that beyond a certain distance from the surface the velocity parallel to the plate remained substantially constant, but as the surface was approached, it began to fall, slowly at first and then more rapidly to become virtually zero at the surface itself (Fig. 1.4a). The region near the surface over which the velocity changes is known as a *boundary layer* and it will be noted that within this layer, the velocity gradient, i.e. the change in velocity with distance, decreases continuously from the surface outwards, from a maximum at the surface to zero at the 'edge' of the boundary layer.

Boundary layers are formed whenever a fluid moves over a surface. In the case of water flow through relatively small tubes for example (Fig. 1.4b), since the water is constrained, the boundary layers that form on the walls meet at the centre; as a result, the velocity of flow increases from zero at the walls to a maximum along the axis. Again we see that the velocity gradient decreases continuously from wall to axis.

To explain how a boundary layer is formed, let us continue with water as the fluid and imagine that under the action of some external driving force, it flows parallel to the surface in the form of layers or streams of molecules. Those molecules in the layer immediately in contact with the surface will tend to adhere to it by intermolecular forces, so that movement of this layer, relative to the surface will be zero. This is known as the condition of '*no slip*' and underlies most of the subsequent arguments on fluid flow over surfaces. Because of the strong attractive forces between water molecules, those in the stationary layer will tend to slow down molecules in the next layer; in turn these will tend to slow down molecules in the layer further away and so on.

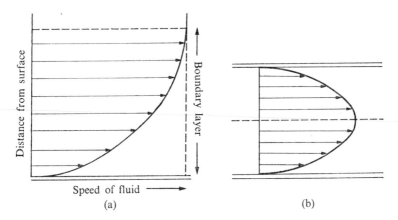

FIG. 1.4. Change in speed of a fluid with distance from (a) the surface of a flat plate and (b) the wall of a circular tube. The boundary layer corresponds to the depth of fluid in which the speed changes with distance from the surface over which it is flowing.

If we make the reasonable assumption that the 'drag' effect originating at the surface decreases with increasing distance from the surface, the net effect is to produce a boundary layer in which the velocity of the fluid increases from the surface until it attains the velocity of the freely moving streams further away.

Gases also produce boundary layers originating from retardation of flow at the surface. However, since the molecules of a gas are more widely dispersed than in a liquid, intermolecular attraction plays a relatively minor role. Instead, we have the phenomenon of '*cross-stream transfer of momentum*'. Each molecule travelling in a stream parallel to the surface (for simplicity let us call this 'horizontal') will have a horizontal momentum corresponding to the product of its mass and horizontal velocity. Because of their thermal energy the molecules of a gas are also in a state of continuous random motion (Brownian movement); as a result, some will cross from one stream to another carrying with them their horizontal momentum. Molecules crossing from slower to faster streams (i.e. away from the surface) will lower the average horizontal momentum and hence the average speed of the latter, whilst molecules crossing from faster to slower streams will have the opposite effect. Thus, cross-stream transfer of momentum, like intermolecular attraction, provides a mechanism for 'transmitting' surface drag through the fluid with the resulting formation of a boundary layer.

A common feature of intermolecular attraction and cross-stream transfer of momentum is the reduction in velocity differences between adjacent streams. In other words, each layer of molecules appears to resist the relative movement of its neighbours as though they were partially 'stuck' to each other. The extent to which this occurs is in fact a measure of the *viscosity* of the fluid; the more viscous the fluid the more difficult it is for the layers to slide over each other, as is readily confirmed when highly viscous liquids are poured or stirred. Viscosity then, can be regarded as a kind of internal friction between layers and if the layers are to be kept moving, the resulting frictional force must be balanced by the external force responsible for propelling the fluid over the surface. Since frictional forces act along the surfaces between layers, the external force must act in the same way and since it tends to shear the fluid it is known as a *shear force*. On a unit surface area basis, this corresponds to a *shear stress* with units of pressure.

It follows from the above that the more viscous the fluid, the greater will the shear stress need to be to maintain a steady flow rate. For a given viscosity, the greater the shear stress, the more the layers will slide over each other and the greater will be the velocity gradient.

The relation between the shear stress, the viscosity and the velocity gradient in a fluid is formally expressed in Newton's law of viscosity:

Shear stress (τ) = viscosity (η) × velocity gradient ($\Delta u/\Delta z$).

More precisely, η is known as the *dynamic viscosity*. The velocity gradient, also known as the *shear rate*, and expressed above as the change in velocity Δu with change in distance from the surface Δz, is usually given in the differential form du/dz. In mathematical form, Newton's law is expressed as:

$$\tau = \eta \frac{du}{dz}. \tag{1.1}$$

Since the shear stress τ has units of a pressure, or force per unit area, it will have the dimensions $ML^{-1}T^{-2}$ (units, $N\ m^{-2}$); the velocity gradient (shear rate) has the dimensions, LT^{-1}/L, i.e. T^{-1} (units, s^{-1}), so that the dynamic viscosity η must have the dimensions $ML^{-1}T^{-1}$ and is expressed in units of $kg\ m^{-1}\ s^{-1}$.

Fluids for which η is independent of the shear stress and shear rate, i.e. those which obey Newton's law, are known as *Newtonian fluids*. For most practical purposes η may be assumed constant within a fluid, in which case the shear stress is proportional to the velocity gradient within the fluid. Since the latter is greatest next to the surface (Fig. 1.4), the shear stress will be a maximum at the surface and is denoted by the symbol τ_0. Since it gives rise to a frictional force, it is commonly referred to as the skin friction, and as we shall see, is the sole source of the drag on a fluid moving in a tube or over a flat surface.

In many problems of fluid motion, the density of the fluid has also to be taken into account and it is then more convenient to express the viscosity as the so called *kinematic viscosity*, v defined by:

$$v = \frac{\eta}{\rho} \tag{1.2}$$

where ρ (kg m^{-3}) is the density of the fluid. The dimensions of v are therefore, $L^2 T^{-1}$ and the (SI) units, m^2 s^{-1}.

Data on the dynamic and kinematic viscosities of water and air at different temperatures, and of sea water are given in Table 1.1.

With both liquids and gases, the dynamic viscosity tends to increase with pressure, but within the range of pressures normally encountered, the change is negligible. The effect of temperature however, is quite different. It will be noted that whereas η for water, as with other liquids, decreases with increasing temperature, that of air and other gases increases. This is explained by the fact that with liquids the main effect of a temperature increase is to decrease the density and the corresponding reduction in the attractive forces between molecules results in a fall in the viscosity. With gases on the other hand, it turns out that with increasing temperature, the decrease in density is less than the increase in the velocity of the molecules and the distance they move, so that the viscosity increases.

The viscosity of solutions and suspensions

Although for the sake of simplicity, much of the subsequent discussion involving the viscosity of liquids refers to pure water, liquids of biological interest almost invariably comprise aqueous solutions of electrolytes and/or organic compounds, or suspensions of particles of varying size. It is, therefore, of some importance to know how these constituents affect the viscosity.

With dilute solutions of electrolytes (c_M below about 0·1), experiments have shown that the viscosity varies with concentration according to the empirical relation:

$$\eta/\eta_0 = 1 + A\sqrt{(c_M)} + Bc_M \tag{1.3}$$

where η/η_0 represents the ratio of the viscosity of the solution η to that of the solvent η_0, and c_M the molar concentration (the number of moles per litre). The coefficient A represents interionic forces and is always positive, whilst B reflects ion-solvent interactions and may be positive or negative depending on species and temperature. Thus, at very low concentrations, with c_M

TABLE 1.1

Physical properties of air and water

Dry air at atmospheric pressure

Temperature θ °C	Density ρ kg m⁻³	Specific heat c_p 10³ J kg⁻¹ °C⁻¹	Dynamic viscosity η 10⁻⁵ kg m⁻¹ s⁻¹	Kinematic viscosity $\nu = \eta/\rho$ 10⁻⁵ m² s⁻¹	Thermal conductivity λ 10⁻² W m⁻¹ °C⁻¹	Prandtl number Pr —	$Gr/l^3\Delta\theta$ $g\beta\rho^2/\eta^2$ 10⁸ m⁻³ °C⁻¹	$Gr\,Pr/l^3\Delta\theta$ $g\beta\rho^2c_p/\eta\lambda$ 10⁸ m⁻³ °C⁻¹
−10	1·34	1·006	1·67	1·24	2·33	0·72	2·44	1·74
0	1·30	1·006	1·72	1·33	2·41	0·72	2·05	1·46
10	1·25	1·006	1·76	1·42	2·49	0·71	1·75	1·24
20	1·20	1·007	1·81	1·51	2·57	0·71	1·47	1·04
30	1·16	1·007	1·86	1·60	2·65	0·71	1·28	0·90
40	1·13	1·008	1·90	1·69	2·72	0·71	1·10	0·78
50	1·09	1·008	1·95	1·78	2·80	0·70	0·94	0·67

Water at moderate pressures

°C	10² kg m⁻³	10³ J kg⁻¹ °C⁻¹	10⁻⁴ kg m⁻¹ s⁻¹	10⁻⁷ m² s⁻¹	W m⁻¹ °C⁻¹	—	10¹⁰ m⁻³ °C⁻¹	10¹⁰ m⁻³ °C⁻¹
0	10·0	4·22	17·9	17·9	0·55	13·7	−0·018	−0·24
10	10·0	4·19	13·1	13·1	0·58	9·5	+0·054	+0·52
20	10·0	4·18	10·1	10·1	0·60	7·0	0·204	1·44
30	10·0	4·18	8·0	8·0	0·62	5·43	0·467	2·64
40	9·9	4·18	6·5	6·6	0·63	4·34	0·88	3·81
50	9·9	4·18	5·49	5·56	0·64	3·57	1·46	5·21
Sea-water 20°C; conc. 0·035 kg kg⁻¹	10·2	3·9	10·9	10·7	0·60	7·1	0·18	1·3

Data from Ede (1967).

below about 0·002, the relationship between viscosity and concentration is curvilinear; at higher concentrations, when the viscosity is dominated by B, it may increase or decrease linearly with concentration. It turns out that within the range of concentrations found in biological systems, the effect of dissolved electrolytes on viscosity is very small. For most common inorganic ions in water, A varies between about 0·005 and 0·02 whilst B is usually below 0·5. Even in the case of sea-water therefore, with a molarity of about 0·6, its viscosity is very little different from that of pure water (Table 1.1).

When the solute molecules (plus solvation shells) are much larger than the solvent molecules, or the liquid contains rigid particles in suspension, increased interference with the pattern of flow increases the viscosity. The effect was first studied by Einstein for dilute suspensions of rigid spheres, for which he established the following relation:

$$\eta/\eta_0 = 1+2\cdot5\,\phi \qquad (1.4)$$

where ϕ represents the fraction of the total volume of suspension occupied by the spheres. Provided that the concentration is low, this equation is valid for suspensions and solutions of non-spherical particles and molecules if the factor 2·5 is replaced by a smaller factor a the value of which depends on the shape (not size) of the particles; in the case of flat discs comparable in shape to red blood cells (p. 168) $a = 2\cdot06$.

An important point concerning elongated particles is their *orientation*. In dilute suspensions, the application of a shear stress tends to orientate otherwise randomly-oriented particles so that they spend most of their time with their long axes roughly parallel to the line of flow. In this position they will interfere least with the flow pattern and will have least effect on the viscosity of the suspension.

As the concentration of the particles increases, they interfere more and more with each other and with the pattern of flow; they will also tend to immobilize an increasing amount of suspending fluid. As a result, the viscosity rises faster than predicted by Einstein's equation above. For rigid spheres, this equation no longer applies when ϕ exceeds about 0·1. Various formulae have been proposed relating the relative viscosity η/η_0 to the molar fraction in concentrated solutions and suspensions, most of them incorporating semi-empirical factors to allow for increased interaction between particles (see Stokes and Mills 1965; Whitmore 1968). However, if the particles are flexible and can change their shape according to flow conditions (e.g. long chain molecules, red blood cells), or if they aggregate or form a continuous structure, the viscosity may no longer be independent of the shear stress. Such fluids are then called *non-Newtonian* and these will be discussed in more detail in Chapter 10.

The development of boundary layers in pipes and over flat surfaces; laminar and turbulent flow

The situation concerning so-called *laminar* flow in pipes is illustrated in Fig. 1.5a. Near the point of entry the velocity across the diameter remains more or less constant, but as the fluid passes through more of the pipe, the viscous drag (skin friction) increases and an increasingly thick boundary layer is formed. At some distance from the mouth of the pipe (see p. 14) the boundary layers meet at the axis and this situation is maintained along the remainder of the pipe.

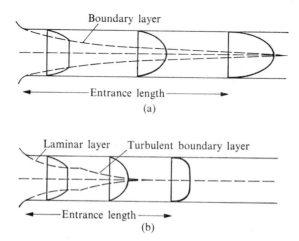

FIG. 1.5. Growth of the boundary layer in a tube with distance from the entry, (a) with laminar flow, (b) with turbulent flow.

The above applies only to relatively low fluid velocities and/or narrow pipes. At higher velocities and/or wider pipes, the nature of the boundary layer and the kind of flow within it change drastically, as was first shown in a series of classical experiments by Reynolds (1883). In order to obtain a visual guide to the pattern of water flow through pipes under different conditions, he introduced a stream of dye into the water (Fig. 1.6). He found that when the flow rate was relatively slow, or the pipe diameter relatively small, the dye appeared as a well defined filament along the axis of the tube (Fig. 1.6a). The different layers of water appeared to follow separate courses parallel to the axis without mixing. This is *laminar flow*. However, when for a given tube diameter, the flowrate was increased beyond a certain critical value, the filament was torn apart and the dye was distributed throughout the water which was now moving very irregularly (Fig. 1.6b). This is called *turbulent flow*.

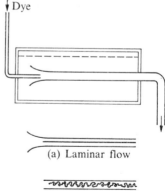

(a) Laminar flow

FIG. 1.6. Diagram of Reynolds' apparatus (1883) for studying flow in tubes showing appearance of dye (a) with laminar flow (b) with turbulent flow.

(b) Turbulent flow

Experiments with different flow rates and different pipe diameters showed that the change from laminar to turbulent flow occurred when the quantity $u_m \rho d/\eta$ exceeded a certain value, of the order of 2000 for smooth pipes (ρ = fluid density, u_m = average velocity of flow, d = internal pipe diameter and η = dynamic viscosity of the fluid). Under particularly stable conditions, it has been established that laminar flow in smooth pipes may persist at much higher values, but turbulent flow does not normally occur at lower values.

The quantity $u_m \rho d/\eta$ (or $u_m d/v$ since $v = \eta/\rho$) is known as the *Reynolds' number*, (*Re*) and inspection of the dimensions of its constituent parameters will show that it is dimensionless, i.e. it is simply a number. Later we shall come across other dimensionless numbers used in the quantitative study of flow problems. As we see in Fig. 1.5b, the shape of the velocity profile across the tube is very different with turbulent flow than with laminar flow; the change in the velocity near the wall is much more rapid (indicating a greater skin friction) but the velocity across the rest of the tube is more uniform.

Laminar and turbulent flow may also be observed when a fluid moves over a flat surface e.g. a thin plate (Fig. 1.7). Let us assume that before it

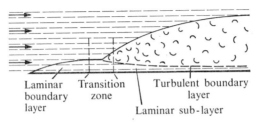

Laminar boundary layer   Transition zone   Turbulent boundary layer

Laminar sub-layer

FIG. 1.7. Formation of laminar and turbulent boundary layers over the surface of a flat plate.

reaches the plate, the mean fluid velocity is uniform. As the fluid passes over the leading edge, then as explained earlier, skin friction slows down the layers of fluid near the surface and a boundary layer is formed. Regardless of the nature of flow in the approaching stream (i.e. whether laminar or turbulent) flow in this boundary layer is initially always laminar. As the fluid passes over a greater area of surface, the total skin friction force increases, flow near the surface becomes increasingly affected and the depth of this laminar boundary layer increases. When a certain length of plate has been traversed, conditions become unstable and flow becomes turbulent. This transition is not immediate but occurs over a finite length of plate known as the *transition zone*.

A significant feature of flow over a plate is that as in the case of pipe flow, transition from laminar to turbulent flow occurs when the quantity $Ul\rho/\eta$ or Ul/ν exceeds a certain critical value. In fact, this quantity is derived in exactly the same way as the Reynolds' number for a pipe, except that a length of plate l is used instead of a pipe diameter d and instead of the average velocity u_m, we use the free stream velocity, U. Depending on the degree of turbulence in the free stream, the critical (Re) for smooth plates ranges from about 5×10^5 with high turbulence to about 3×10^6 with low turbulence; in other words, the more turbulent the approaching flow, the earlier is turbulence initiated over a plate.

As still more of the plate is traversed, the turbulent boundary layer increases in depth, but below this, in the region immediately next to the surface, there persists a thin layer of fluid in which flow remains laminar. This is known as the *laminar sub-boundary layer* and it decreases in depth as the turbulent boundary layer increases. A similar laminar sub-boundary layer occurs below the turbulent boundary layer in pipes.

Wind flow over the ground is almost invariably turbulent, except of course, for the laminar sub-layer next to the surface. Nevertheless, it will be appreciated that when wind (or turbulent water) passes over a small discrete surface e.g. that of a leaf or animal body, flow over these surfaces may still be laminar. The effect can readily be seen in streams when otherwise turbulent flow becomes much smoother and laminar when it passes over a flat rock near the surface.

The type of fluid flow over surfaces, whether laminar or turbulent, is of considerable biological importance. In laminar flow the properties of the fluid can only be transported across stream by the random thermal movement of the molecules, i.e. by molecular diffusion. This means that the rate of transport of heat, gases, and dissolved substances to and from a surface will be relatively slow, especially in liquids (p. 118). When turbulent flow occurs, the situation is very different. It is customary to picture turbulence as the random movement of discrete parcels or *eddies* of fluid from one stream to another, thus providing a very efficient transport mechanism within the fluid, known as *eddy diffusion* or *turbulent transport*, for momentum, heat and other fluid properties. As an example, we might take the commonly-observed fluttering of leaves on a tree in wind. If a record is made of the horizontal wind speed at some height above the ground, it will usually be found to fluctuate irregularly with time (Fig. 1.8). A mean value can be established from such a record

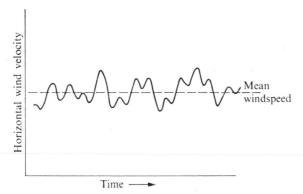

FIG. 1.8. Hypothetical variation in horizontal wind velocity above the ground with time.

if sufficient time has been allowed to provide a representative sample of the fluctuations. These fluctuations arise from the random cross-stream transport of momentum by the eddies. Temporary increases in the wind-speed (*gusts*) are due to the arrival of eddies from faster-moving air streams above and temporary drops in wind-speed (*lulls*) from the arrival of eddies from slower-moving air streams nearer the ground.

Later we shall need to go into these transport mechanisms in more detail; for the present it will be sufficient to recognize in general terms, the essential differences between laminar and turbulent flow in boundary layers.

2. Flow in tubes

HAVING dealt with the general principles underlying the nature and formation of boundary layers, we can now turn to some of the quantitative relationships that are of importance in biological systems. As explained in the
Preface, most of the details concerning the derivation of these relationships
are given in Appendices, but readers are encouraged to try to follow them,
not only to improve their understanding of the principles involved, but also
to help them to appreciate the limitations imposed by the various assumptions that are made.

In the systems with which we shall be concerned, flow results essentially
from the application of a pressure or tension (i.e. suction) to the fluid; because
of skin friction (τ_0) opposing the movement of the fluid, energy must be expended to keep the fluid moving and as a result, there will be a fall in pressure
along the tube. The fall in pressure per unit length of tube is known as the
pressure gradient; if along a length of tube l, the pressure drop is Δp, the
pressure gradient will be $\Delta p/l$. If the flow rate is constant, it can then be
shown (Appendix 1) that the shear stress τ at a distance y from the axis is
given by:

$$\tau = \frac{\Delta p y}{2l} \tag{2.1a}$$

and that at the wall (skin friction τ_0) where $y =$ tube radius, r by:

$$\tau_0 = \frac{\Delta p r}{2l}. \tag{2.1b}$$

In other words, the pressure gradient $\Delta p/l$ is directly proportional to the
skin friction and inversely proportional to the tube radius. This applies to
both laminar and turbulent flow.

Laminar flow

For a fully-developed laminar boundary layer an entry length is required
corresponding to about $0.06(Re)d$ (where $d =$ diameter) (Fig. 1.5); at the
critical (Re) of 2000 therefore, this would amount to about 120 times the
diameter.

For straight circular tubes with rigid walls and a constant flow rate, the velocity profile is then parabolic in form (Fig. 1.5a and Appendix 1) and the volume flow rate q_v (m³ s⁻¹) is given by:

$$q_v = \frac{\pi \Delta p r^4}{8 \eta l}.$$ (2.2)

This equation is known as the Hagen–Poiseuille (or simply Poiseuille) law and shows that for a given pressure gradient, $\Delta p / l$, the volume flow rate, q_v increases as the 4th power of the tube radius r and inversely as the dynamic viscosity η of the fluid. Under the same pressure gradient therefore, a tube of say, 1 mm radius will conduct fluid at a rate 10 000 times faster than a tube of 0·1 mm radius. As we shall see later, this law has been widely applied to biological systems where it is necessary to relate flow rates to pressure gradients and tube dimensions.

As shown in Appendix 1, the maximum velocity of flow occurs along the tube axis and is given by:

$$u_{max} = \frac{\Delta p r^2}{4 \eta l}.$$ (2.3)

If, instead of the maximum we wish to use the average velocity u_m across the whole cross sectional area of the tube ($= \pi r^2$), this is easily obtained from eqn (2.2) by substituting $q_v = \pi r^2 u_m$, whence;

$$u_m = \frac{\Delta p r^2}{8 \eta l}.$$ (2.4)

It will be noted that with laminar flow the average velocity u_m is exactly half the maximum velocity u_{max}.

The capillary tube viscometer

The capillary tube viscometer, widely used for measuring the viscosity of liquids, is based on Poiseuille's law in the form:

$$\eta = \frac{\pi \Delta p d^4}{128 q_v l}$$

where d is the tube diameter. In the Ostwald type of viscometer (Fig. 2.1), the driving pressure, Δp is the hydraulic head of the fluid itself in the bulb AB. The viscosity is then determined from the time taken for the fluid level to fall from A to B and the dimensions of the capillary. Since the hydrostatic head decreases as the bulb empties, this particular instrument must be calibrated (usually in terms of water) and is only really suited to Newtonian liquids in which variations in pressure, i.e. in shear stress, have no effect on

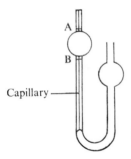

FIG 2.1. Ostwald capillary viscometer (for explanation see text p. 15).

the viscosity. Nevertheless, this instrument has been widely used for non-Newtonian fluids like blood, though in recent years it has been replaced by other types of viscometer (see p. 169). In practice, corrections must be made for flow pattern changes and anomalous pressure gradients at the inlet and outlet of the tube.

Energy relations

Since fluid propulsion is an energy consuming process, two further quantities will be of interest to the biologist, the total force resisting fluid motion and the rate at which energy must be expended (i.e. the power) to keep the fluid moving.

The resistive force per unit area of wall, τ_0 (N m^{-2}) is given by eqn (2.1b). For a length of tube l(m) with a total surface area of $2\pi r l$(m^2) the total resistive force F will then be:

$$F = 2\pi r l \tau_0 = \pi r^2 \Delta p \quad \text{(N)} \qquad (2.5)$$

Alternatively, by substituting for Δp from eqn (2.4), we can relate τ_0 and F to the average velocity u_m, thus:

$$\tau_0 = \frac{4\eta u_m}{r} \quad \text{and} \quad F = 8\pi \eta l u_m \quad \text{(N)}. \qquad (2.6)$$

The power P (watts) required to maintain flow is most simply derived from the product of the resistive F and the average velocity u_m. From eqn (2.5) therefore,

$$P = \pi r^2 u_m \Delta p = A u_m \Delta p, \qquad (2.7a)$$

where A is the pipe cross sectional area.
From eqn (2.6),

$$P = 8\pi \eta l u_m^2. \qquad (2.7b)$$

The power consumption therefore, increases as the square of the average velocity.

Food transport in plants

To illustrate the application of the relationships established above for laminar flow, we might take as our first example, the transport of organic foodstuffs in plants, which, it is generally agreed, takes place in specialized elements of the phloem tissue, the sieve tubes. These consist of long tubular cells, typically 20 to 30 μm in diameter and 100 to 500 μm long, joined end to end through cross walls, the so-called sieve plates (Fig. 2.2). The sieve plates are perforated by pores, about 1 μm long and 0·1 to 5 μm in diameter, and occupy about half the sieve plate area.

The material transported is mainly sucrose at concentrations of about 10 to 30% by volume and at rates ranging from about 2 to as high as 25 g per cm² phloem tissue per hour, corresponding roughly to velocities of about 0·5 to 2 m per hour. These rates are far too high to be accounted for by molecular diffusion, especially since the sieve tubes only occupy about 20 per cent of the area of phloem tissue. The general conclusion is that the transport process must involve some form of mass flow or activated flow, but how this is effected is still the subject of much debate.

One of the most popular and longstanding theories is the pressure flow hypothesis of Münch, according to which the phloem contents are propelled along the sieve tubes by a hydrostatic (turgor) pressure gradient. This is assumed to derive from the osmotic intake of water at the source (e.g. the leaves) where the sugar concentration is high and the release of water at the sinks (regions where the sugar is metabolized) where the concentration is low.

Attempts have been made to apply Poiseuille's law to calculate the pressure gradients required for a given rate of flow. According to the dimensions of the sieve tubes and estimated flow velocities, only laminar flow may be assumed. Using eqn (2.4) for a sieve tube radius of 12 μm (willow) (see Fig. 2.3), Weatherley and Johnson (1968) calculated that a pressure gradient of about $2·5 \times 10^4$ N m^{-2} ($\approx$0·25 bars) per m would be required to propel a 10% solution of sucrose ($\eta = 1·5 \times 10^{-3}$ kg m^{-1} s^{-1}) through the sieve tube lumina at a velocity of $2·8 \times 10^{-4}$ m s^{-1} (1 m h^{-1}): this is equivalent to a mass transport rate of 10 g sucrose per hour per cm² sieve tube area.

To allow for the resistance of the sieve plates, it was assumed that these are occupied by pores of radius 2·5 μm and length 5 μm and that they occur at a frequency of 6 sieve plates per mm sieve tube length. Assuming that Poiseuille flow still applies, the same equation was used to show that the sieve plates would account for an extra pressure gradient of about $3·2 \times 10^4$ N m^{-2} (about 0·32 bar) per metre. Thus the total pressure gradient required would be about 6×10^4 N m^{-2} ($\approx$0·6 bar) per metre.

Measured osmotic pressures of phloem exudate, including the contributions of solutes other than sucrose, suggested a maximum of about 15×10^{-5} N m² (about 15 bars); on this basis it was concluded that the phloem contents could only be transported by pressure flow over a distance of about 25 m, i.e. much less than the height of the tallest trees. Furthermore, if it is accepted that the pores are partially occluded by filamentous material, as suggested by electron micrographs, the pressure gradients would need to be higher still.

Canny (1973) using the same approach, but assuming a sieve plate frequency of 2 per mm, calculated for different sieve pore sizes, the pressure gradients required to propel the same 10% sucrose solution through the sieve pores at a rate of 2 m h^{-1} (= 1 m h^{-1} through the sieve tubes assuming the pores occupy 50 per cent of the sieve plate area). His results (Fig. 2.3)

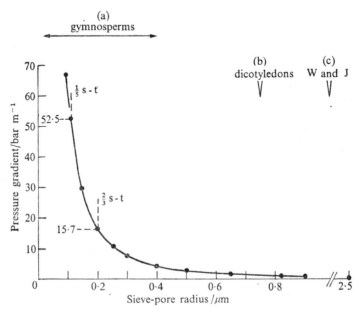

FIG. 2.3. Pressure gradients required to propel a 10% sucrose solution through sieve tubes at a speed of 1 m h^{-1} for various sieve pore sizes. Average radii of gymnosperm and angiosperm pores are shown, together with the pore size assumed by Weatherley and Johnson (1968). From Canny (1973).

emphasise the importance of sieve pore size in distinguishing between the pressure gradients required for angiosperms (relatively large pores) and gymnosperms (relatively small pores).

Another approach to the problem is provided by the energy relations of transport (Spanner 1962; Weatherley and Johnson 1968; Canny 1973). Assuming that the energy released by the respiration of the phloem tissue is available for sieve tube transport, eqn (2.7a) can be modified to allow for the calculation of the corresponding pressure gradient:

$$P = Alu_m \Delta p/l$$

where P = rate of energy consumption (watts), A = cross sectional area of sieve tube lumen (m^2), u_m = average velocity of flow through the lumen (m s^{-1}), l = length of tissue (m) and $\Delta p/l$ the pressure gradient (N m^{-2} m^{-1}).

(In practice smaller units would be more realistic, but for the sake of consistency, SI units are retained.)

For unit area ($A = 1$ m^2) and unit length ($l = 1$ m), corresponding to 1 m^3 volume of sieve tube, $Al = 1$ and the above eqn reduces to:

$$P = u_m \Delta p/l \quad \text{(W)}.$$

Canny (1973) bases his calculations on an energy production of 580 W per m^3 phloem tissue (assuming unit density); if the sieve tubes occupy $\frac{2}{3}$rds of the phloem, the maximum energy available per m^3 sieve tube will be about 870 W. Substituting this for P, with $u_m = 5 \cdot 6 \times 10^{-4}$ m s^{-1} ($= 2$ m h^{-1}) gives a pressure gradient $\Delta p/l$ of 15·6 bar m^{-1}. If the sieve tubes only occupied 20 per cent of the phloem tissue, more energy would be available to maintain a higher gradient of about 52 bar m^{-1}. These values are entered in Fig. 2.3 to show the relation with sieve pore size.

It is not the intention here to discuss the significance of these calculations in relation to the mechanism of phloem transport, the main purpose being to show how the principles of fluid mechanics can be applied to problems of this kind. Further reference to this problem will be found in Chapter 11.

Turbulent flow

We have already seen that turbulent flow in straight circular tubes with smooth walls is initiated when the Reynolds' number exceeds a critical value of about 2000. Generally speaking, a fully turbulent boundary layer is not developed until at least 50 diameters from the entrance to the tube (Fig. 1.5b), but the exact value varies with flow conditions.

It would appear from the order of the sizes of the conducting elements and the velocities of flow in biological systems that the Reynolds' number would be far too low for turbulence to occur. For example, in a smooth tube 10 mm in diameter ($d = 0 \cdot 01$ m), water ($\nu = 10^{-6}$ m^2 s^{-1}) would have to flow at an average speed of at least 0·2 m s^{-1}, and air ($\nu = 1 \cdot 5 \times 10^{-5}$ m^2 s^{-1}) about 7 times faster, for turbulence to be initiated. However, the generally-accepted critical Reynolds' number only applies to straight tubes of circular cross-section. If the tube is curved it seems that turbulence can occur at lower numbers, and the greater the degree of curvature, the lower the critical value. Using glass S, T, and Y shaped models to represent cerebral arteries, Stehbens (1960) found turbulent flow to occur at well below the critical (Re) of 2000. Appreciable turbulence has been observed in the heart, the aorta, and in the trachea, and transient turbulence is common wherever sharp bends or branches occur in the transport system. Local eddies, indicative of turbulent conditions, almost always occur where there is an abrupt change in the size of the conducting channel, as when a fluid is forced to flow through small orifices. These eddies may not persist for any distance, but in so far as they consume energy, they are of some biological importance.

3

Since turbulence is basically a random phenomenon, not yet completely understood, no complete analytical solutions are available for flow in the turbulent boundary layer. We shall not concern ourselves with the more theoretical aspects since it will be sufficient for our purposes to use the quantitative data obtained more or less empirically from experiments.

We might begin therefore with measurements made on the velocity of flow in pipes at different distances from the wall. Because of the fluctuating nature of this velocity within a turbulent boundary layer, it will be appreciated that we are dealing with average values over a period of time (see Fig. 1.8).

Besides the velocity of flow (u) and the distance from the wall (z), the other variables that have to be included are the density of the fluid (ρ) and its kinematic viscosity (ν). To simplify the presentation of the results these four variables are grouped into two dimensionless quantities both incorporating a so-called *friction velocity*, u_* defined as:

$$u_* = \sqrt{(\tau_0/\rho)} \tag{2.8}$$

where, as before, τ_0 refers to the shear stress at the wall (skin friction). It can easily be confirmed that u_* has the dimensions of a velocity.

The two dimensionless quantities are a so-called *reference velocity* u/u_* and a *reference length* zu_*/ν. Fig. 2.4 illustrates the results obtained when

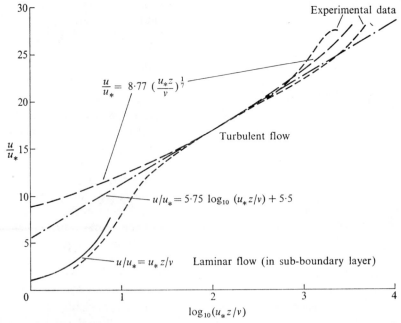

FIG. 2.4 Plot of u/u_* against $\log_{10}(zu_*/\nu)$ for turbulent flow in a pipe, with equations fitted to different regions of the curve; u = mean velocity of flow parallel to the axis at a distance z from the wall and u_* = friction velocity = $\sqrt{(\tau_0/\rho)}$.

these are plotted against each other on a linear–log scale for turbulent flow in smooth pipes over a wide range of flow conditions. Except at very high values $(zu_*/v > 1000)$, almost identical results have been obtained for turbulent flow over a flat plate (p. 39).

Near the wall a good fit to the curve is given by the equation:

$$\frac{u}{u_*} = \frac{zu_*}{v} \quad \text{or} \quad \frac{uv}{z} = u_*^2 = \frac{\tau_0}{\rho} \tag{2.9}$$

whence,

$$\tau_0 = \rho v \frac{u}{z} = \eta \frac{u}{z}. \tag{2.10}$$

Since eqn (2.10) is formally identical with eqn (1.1), it follows that flow near the wall is laminar, thus confirming the occurrence of a laminar sub-boundary layer between the wall and the turbulent zone beyond. The depth of this layer (δ_s) is then given by the distance z over which eqn (2.9) applies.

Since there is no sharp boundary between the laminar and turbulent zones, no precise depth can be specified. Some authors assume that eqn (2.9) only applies up to a value of 5, below which flow is definitely laminar, whereas others extend this value to 11·5. A useful working value is 10, in which case (see Appendix 3), we obtain:

$$\delta_s = \frac{50d}{(Re)^{\frac{7}{8}}} \tag{2.11}$$

where $d =$ tube diameter. In other words, the depth of the laminar sub-layer decreases almost inversely as the Reynolds' number.

In the case of water $(v = 10^{-6} \text{ m}^2 \text{ s}^{-1})$ flowing through a smooth pipe of diameter 0·01 m with an average velocity $u_m = 0.25 \text{ m s}^{-1}$, $(Re) = 2500$, just high enough to produce turbulent flow. The depth of the laminar sub-layer will then be about 0·53 mm.

For the turbulent zone proper, various empirical equations have been proposed to fit different regions of the experimental curve (Fig. 2.4).

Above about $zu_*/v = 100$, a reasonably good fit is given by:

$$\frac{u}{u_*} = 8.77 \left(\frac{zu_*}{v}\right)^{\frac{1}{7}}. \tag{2.12}$$

It can be shown from this equation (Appendix 2) that u_m, the average velocity of flow over the whole tube cross-section, is about 0·8 times the maximum flow rate along the axis, and that the shear stress at the wall (τ_0) is:

$$\tau_0 = 0.0791 \frac{\rho u_m^2}{2} \left(\frac{v}{u_m d}\right)^{\frac{1}{4}}$$

or,

$$\tau_0 = 0.0791 \frac{\rho u_m^2}{2} (Re)^{-\frac{1}{4}} \tag{2.13}$$

where as before, d = tube diameter and $(Re) = \dfrac{u_m d}{v}$.

By substituting for τ_0 from eqn (2.1b) and putting q_v, the volume rate of flow = $\pi r^2 u_m$, it turns out (Appendix 2) that unlike Poiseuille's law, Δp is approximately proportional to q_v^2 and inversely proportional to r^5. We shall use these relationships later when discussing resistances to breathing.

A comparison between eqn (2.13) and eqn (2.6) shows that whereas with laminar flow, the skin friction (τ_0) is directly proportional to the average flow velocity (u_m), with turbulent flow, it is proportional to $u_m^{\frac{7}{4}}$, very nearly u_m^2. This means that much more energy must be expended to maintain turbulent flow at the same average velocity. As we shall see later (p. 25) the difference is further accentuated when the walls of the pipe are rough.

Resistances to breathing

It is appropriate to describe here some of the measurements that have been made to determine the resistance to the movement of air in the human respiratory system during breathing. Clearly a knowledge of this is of considerable physiological importance.

Although laminar flow usually prevails, some turbulent flow occurs in the nose and larynx and especially in the trachea, so that account has to be taken of the resistances to both types of flow. A useful empirical way of dealing with this situation, also used for blood flow, is to look upon the pressure difference established across the whole system (Δp) as that required to produce a volume rate of air flow q_v against two resistances in series with each other, thus:

$$\Delta p = k_1 q_v + k_2 q_v^2$$

where k_1 refers to laminar, and k_2 to turbulent flow. In the case of laminar flow, we can readily rearrange Poiseuille's equation (eqn 2.2) in the form

$\Delta p = k_1 q_v$ where $k_1 = \dfrac{8\eta l}{\pi r^4}$; for turbulent flow (Appendix 2), Δp is approxi-

mately proportional to q_v^2. In experimental practice, a simultaneous record is made of the pressure difference between the mouth or nose and the alveoli of the lungs (giving Δp) and the volume rate of air flow during inhalation and exhalation (giving q_v).

Traditionally, Δp is expressed in units of cm water (1 cm water = 100 N m^{-2}) and q_v as dV/dt or simply $\dot{V}$ litres per second, where V corresponds to the volume of air present in the respiratory system at any given time (1 l s^{-1} = 10^{-3} m^3 s^{-1}). The resistance coefficients, k_1 and k_2 are then expressed in physiological units of cm water per litre per second. By plotting $\Delta p/q_v$ against q_v, k_1 can be obtained from the intercept and k_2 from the slope.

Otis, Fenn, and Rahn (1950) have extended this type of study to determine the total mechanical work done by the respiratory muscles during inhalation. Besides overcoming frictional resistance to air flow, energy must also be expended to overcome other resistive forces, the most important of which are the elastic forces opposing the expansion of the chest and lungs. Since the pressure necessary to overcome these elastic forces will be approximately proportional to the volume present, a third component kV can be introduced into the above equation to account for this:

$$\text{Total pressure required} = kV + k_1\dot{V} + k_2\dot{V}^2$$

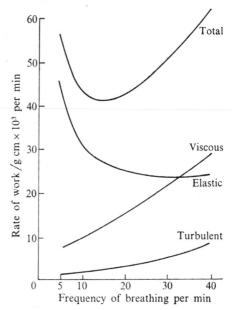

FIG. 2.5. Elastic, viscous, turbulent, and total work as a function of breathing frequency in man when alveolar ventilation $= 6 \, \text{l min}^{-1}$ and dead space $= 0.21$. After Otis *et al.* (1950).

On the basis of this equation, the authors have been able to work out experimentally the contributions of the three sources of resistance to the muscular work done at different breathing rates. The significant feature of their results (Fig. 2.5) is that the frequency of 15 breaths per minute corresponding to a minimum work rate lies within the range of the natural breathing frequency of resting subjects. A lower frequency would mean the expenditure of too much energy in moving large volumes of air against the elastic resistance of chest and lungs, whilst at a higher frequency, energy would be wasted in ventilating dead space in the respiratory system.

Pipe friction coefficients

Because of its practical importance, engineers have paid considerable attention to the problem of skin friction, especially where turbulent flow is concerned. As we have already seen from the example described above, the biologist is also concerned with this problem and it is therefore of some interest to take a closer look at the methods adopted by engineers.

The common custom is to relate the skin friction τ_0 to a dimensionless number, the so-called pipe friction coefficient C_t defined as:

$$C_t = \frac{\tau_0}{\frac{1}{2}\rho u_m^2} \qquad \text{whence} \qquad \tau_0 = C_t \tfrac{1}{2}\rho u_m^2, \qquad (2.14)$$

where ρ and u_m refer to the density and mean velocity of the fluid respectively. This scheme considerably simplifies the presentation of experimental data since instead of several parameters, we have a single number, C_t characteristic of a particular set of flow conditions. For example, in the case of laminar flow, it can be readily shown from eqn (2.6) that $C_t = 16\nu/u_m d$ and since $u_m d/\nu$ corresponds to the pipe Reynolds' number (Re), we obtain the simple relation:

$$C_t = \frac{16}{(Re)}. \qquad (2.15)$$

For turbulent flow in smooth pipes, we can rewrite eqn (2.13) as:

$$C_t = 0\cdot0791\,(Re)^{-\frac{1}{4}}. \qquad (2.16)$$

In both cases therefore, C_t can be presented as a sole function of the Reynolds' number, as illustrated in Fig. 2.6. Apart from the Reynolds' number, all that is needed to calculate C_t for any particular set of conditions is the fluid density and the average velocity of flow.

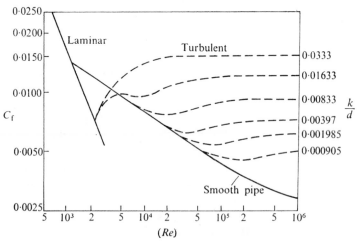

FIG. 2.6. Pipe friction coefficient C_t as a function of the Reynolds' number (Re) and wall roughness ratio k/d.

The main purpose of Fig. 2.6 however, is to illustrate the effect of wall roughness on resistance to flow. So far we have only dealt with smooth pipes, though without defining what this means. Theory, supported by experiment, has shown that if the roughness of a surface is expressed as the average height k of the surface irregularities, or 'bumps', then C_f can be expressed as a function of both (Re) and the ratio k/d, where d is the diameter of the pipe. The data shown in Fig. 2.6 were obtained by gluing variously sized sand grains to the walls of pipes and then determining C_f from the pressure gradient and rate of volume flow under a variety of flow conditions.

It will be seen that below about $(Re) = 2000$, $C_f = 16/(Re)$, thus confirming eqn (2.15); with laminar flow therefore, pipe friction is independent of wall roughness. This is of some biological interest since the walls of a number of natural conducting systems are often irregular. By contrast, at higher (Re), when flow becomes turbulent, C_f increases with wall roughness (increasing k/d) and only when the wall is smooth ($k/d = 0$) does the curve correspond to eqn (2.16).

An interesting feature of these curves is that for any given value of k/d, C_f attains a constant value at a certain (Re) and that this value of (Re) decreases as k/d increases, as shown by the dashed lines. The explanation for this rests on the supposition that when the roughness elements are small enough to be submerged in the laminar sub-layer that lies below the turbulent layer, they have little or no effect on the drag. The larger the k/d ratio, the more they project into the turbulent zone, the more local eddies they create and at least up to a certain point, the higher will be C_f. Since the depth of the laminar sub-layer decreases with increasing (Re) (eqn 2.11), it follows that 'bumps' which may be submerged at low (Re) and therefore, do not affect C_f, may project beyond it at higher (Re) and then increase C_f. In other words, a surface may behave as smooth at low (Re) (low velocity) but rough at higher (Re) (higher velocity).

The above is a simplified explanation of what is actually a rather complex phenomenon. There is no sharp boundary between the laminar sub-layer and the turbulent zone above, and the simple concept of an average height of the roughness elements makes no allowances for other factors such as their shape and spacement.

Warning Most authors refer to the pipe friction coefficient simply as f and usually this is identical with the C_f defined above. According to some authors however, $f = 4C_f$ so that it is always necessary to check when reading particular texts. A more serious source of confusion, especially in older texts, is the custom of expressing τ_0 and other pressures in terms of say, kg (weight) per unit area. As explained earlier (p. xi) a force of 1 kg weight $= 9.81$ newtons, but the conversion is not always made clear. Readers are strongly advised to check all texts which have not yet adopted the S.I. system of units.

Departures from the ideal

It should be remembered that the quantitative relationships established above strictly apply only to Newtonian fluids and straight circular tubes with rigid walls, and of sufficient length for the flow profile to be fully developed. Probably few biological conducting systems conform exactly to these conditions. If we are to assess the validity of calculations based on the above formulae, we must, therefore, consider the effects of departures from the ideal.

Mention has already been made of the lowering of the critical Reynolds' number when the conducting elements are curved, but it appears that the

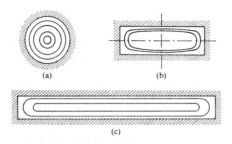

FIG. 2.7. Contours of equal flow velocity in pipes of different shapes.

effect is small unless the curvature is of the same order as the tube radius. Changes in the size of the conducting elements will also introduce errors, e.g. it will be remembered that according to Poiseuille's law, the volume flow rate is proportional to the fourth power of the radius. If the change in size is abrupt, local eddying will certainly involve a greater expenditure of energy.

When the conducting elements are not circular in cross section, flow is no longer symmetrical about the axis nor the drag forces uniform around the perimeter. As illustrated in Fig. 2.7, the contour lines of equal velocity in a rectangular pipe are more crowded near the mid-points of the walls than near the corners. Since the skin friction is proportional to the velocity gradient, this means a greater drag force at the mid-points than elsewhere. For a somewhat square section like (b) corresponding to most conifer water-conducting elements (Fig. 3.3) the corners influence the flow much more than in a long narrow section like (c). On the other hand, the difference in velocity gradients between the shorter and longer sides of (c) is greater than that in (b). To some extent therefore, these effects compensate each other and it has been established that the pipe friction coefficients are not so different from that of circular pipes as usually to introduce serious errors.

The effect of distensible walls such as occur in blood vessels, sieve tubes and latex vessels, is to make the size of the vessel dependent on the driving pressure, so altering the relation between the pressure and flow rate. Although it is not difficult to picture this effect in general terms, the mathematics are

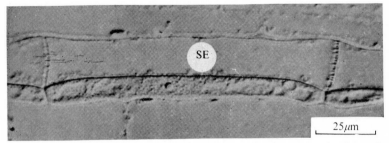

FIG. 2.2. Sieve element (SE) of *Nicotiana tabacum* showing pores in sieve plates. From Cronshaw and Anderson (1969).

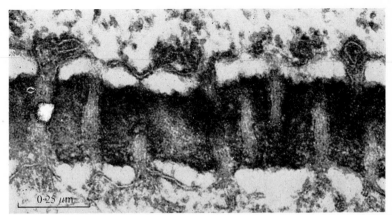

FIG. 2.8. Plasmodesmata in leaf gland of *Drosera capensis*. Electronmicrograph by A. I. Gilchrist (School of Botany, Oxford).

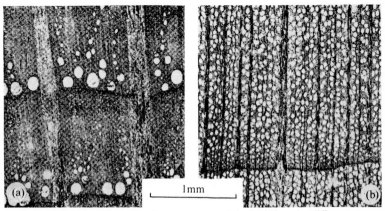

FIG. 3.5. Transverse sections of a ring-porous wood (oak, (a)) and a diffuse-porous wood (beech, (b)).

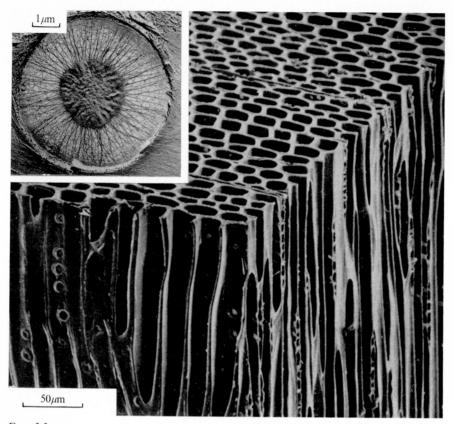

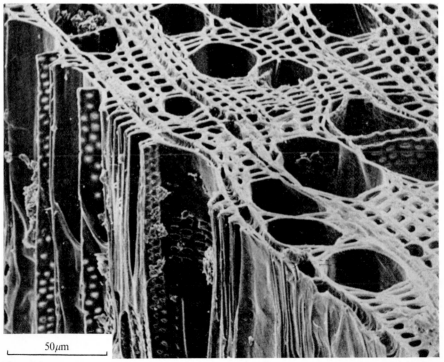

FIG. 3.3.

FIG. 3.4.

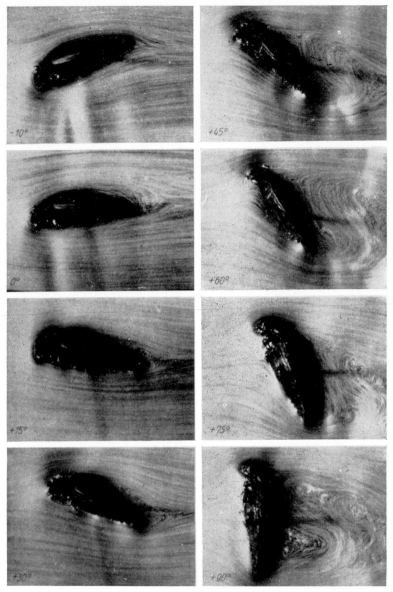

FIG. 5.7. Streamlines around the body of the water-beetle *Dytiscus* in water flowing at a speed of 10 m s^{-1}. Note increasing turbulence as the angle of the body to the direction of water flow increases. From Nachtigall and Bilo (1965).

FIG. 3.3 Scanning electronmicrograph of conifer wood (*Pinus radiata*). Note bordered pits connecting tracheids. From Meylan and Butterfield (1971). Inset: electron-micrograph of unaspirated bordered pit showing torus and membrane (from Petty, 1972) with permission of the Royal Society.

FIG. 3.4. Scanning electronmicrograph of angiosperm wood (red beech, *Nothofagus fusca*). Note vessels, fibres, ray parenchyma cells with simple pits between vessels. From Meylan and Butterfield (1971).

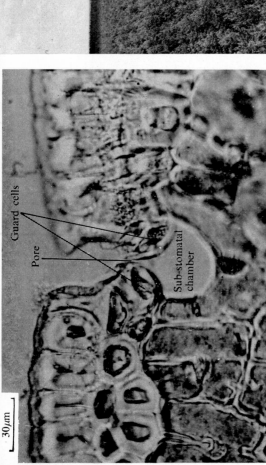

FIG. 8.1. Transverse section of a needle of *Pinus nigra var. austriaca* showing stomatal pore sunk below the surface of the epidermis. Photograph by M. Nikolaides.

Guard cells

Pore

Sub-stomatal chamber

30μm

FIG. 9.9. Instruments mounted on a mast for taking measurements of air temperature, humidity etc. at different levels above a pine forest at Thetford, Norfolk. Photograph by courtesy of the Natural Environment Research Council Institute of Hydrology, Wallingford, Berks.

too complicated to discuss here. In the case of blood, the situation is further complicated by its pulsatory nature; this will be discussed later (p. 174).

Flow through narrow tubes

Normal flow relations may be affected if the tubes are very small. If the liquid contains a suspension of particles, anomalous flow often occurs, especially if the size of the particles is of the same order as that of the tube. Some of these particulate effects have already been discussed (p. 9) and more will be said later (p. 172). However, even in the case of water and aqueous solutions, complications may arise if the tubes are extremely small.

As described earlier (p. 3), hydrophilic surfaces are probably covered by a layer of ordered water with very different properties than the bulk water further away. It has a much lower solubility for salts and according to Schultz and Asunmaa (1970), a viscosity about 35 times higher than ordinary water at the same temperature.

When the tubes are relatively large, this surface phenomenon will have negligible influence on water flow, but when, as in many natural and artificial membranes, the pores are minute, the sheath of ordered water lining the tubes may substantially reduce the permeability of the membrane for water and solutes. In fact, according to Schultz and Asunmaa, pores with a diameter of 4·4 nm or less, will be completely filled with ordered water and will then act as a solute filter; this they suggest, explains the use of cellulose acetate and porous glass as desalination membranes. They found excellent agreement between measured permeabilities of cellulose acetate membranes with a mean pore radius of 2·1 nm and those calculated from Poiseuille's equation when allowance was made for the higher viscosity of the water.

Clarkson, Richards, and Sanderson (1971) among others, have used Poiseuille's law to estimate the pressure gradients required to propel water at a given rate through plasmodesmata. These appear as minute tubules lined with plasmalemma crossing the walls of adjoining cells in plant tissues (Fig. 2.8). Some plasmodesmata at least, appear to have even smaller tubules (desmotubules) running along the axis and apparently connecting the endoplasmic reticulum of the cells. By maintaining cytoplasmic continuity between cells (the so-called symplasm), plasmodesmata are believed to provide an important pathway for the conduction of water and solutes across tissues. This is particularly important in transport from the cortex to the stele in roots because the intervening endodermis has a fatty deposit in the cell walls (the Casparian strip) which precludes transport through the walls.

From electron micrographs of the tertiary endodermis of barley roots, Clarkson *et al.* estimated the radius of the plasmodesmata tubules and of the desmotubules to be 10 nm and 5 nm respectively, thus only about 5 times and twice the presumed thickness of the layer of ordered water. Probably the error is not too great in the case of transport through the plasmodesmata if the whole of the tubule is available for flow, but this appears very doubtful in the case of desmotubule transport. Using a similar approach, Tanton and Crowdy (1972) concluded that under an assumed pressure gradient of 5×10^5 N m^{-2}, the plasmodesmata would transport about 22 per cent of the measured water flow if they were open tubes and only about 2 per cent if the cross sectional area was reduced by desmotubules. They concluded therefore, that the symplastic pathway plays a minor role in transfer across the root.

Slip flow

It has long been known that if the pressure is reduced, the rate of flow of a gas through a capillary exceeds that calculated by Poiseuille's law. This phenomenon is explained by the increased mean free path of the molecules, i.e. the average distance travelled before colliding with other molecules.

If the pressure is low enough, and/or the capillary small enough so that the mean free path is much greater than the capillary diameter, the gas molecules will tend to collide with the walls rather than with each other. In the absence of intermolecular transfer of momentum (p. 5), viscous flow, on which Poiseuille's law is based, no longer occurs; instead there is a flow or 'drift' of free molecules along the pressure gradient in the tube. This is known as 'slip' or *Knudsen flow*.

The mean free path λ of a molecule may be calculated from:

$$\lambda = \frac{2\eta}{\bar{p}} \sqrt{\left/ \left(\frac{RT}{M}\right)\right.} \tag{2.18}$$

where η is the dynamic viscosity, $\bar{p}$ the average pressure, T the absolute temperature, M the molecular weight of the gas, and R the universal gas constant (8·31 J mol^{-1} K^{-1}).

For air at a temperature of 20°C ($T = 293$ K) and at atmospheric pressure ($\bar{p} = 10^5$ N m^{-2}), with $\eta = 1 \cdot 81 \times 10^{-5}$ kg m^{-1} s^{-1} and $M = 0 \cdot 029$ kg mol^{-1}, the mean free path turns out to be 10^{-7} m or 0·1 μm. Capillary diameters of this size or even smaller are not uncommon in porous media so that even at ordinary pressures, slip flow would be expected to predominate. In larger capillaries some slip flow will occur within a distance λ from the wall where the molecules can reach the wall without colliding with each other. The effect of course, becomes relatively smaller with increasing size of the capillary and above a radius of about 100 λ, flow can usually be treated as entirely viscous.

Within the transition range in which both viscous and slip flow occur, the two are added together. However, except for rather special applications such as in the study of the conducting properties of wood (p. 37) and other porous media, this phenomena is perhaps only of marginal biological interest. Detailed accounts are given in several texts including Carman (1956) and Dushman (1962).

3. Flow in porous media

ALTHOUGH, as we have seen, there are certain biological conducting systems that can be treated in terms of flow through individual tubes (blood flow will be dealt with in Chapter 10), most systems correspond to porous media in which flow occurs through an assembly of channels, often interconnecting and varying in size and shape. Examples include membranes, plant cell walls, wood, and soils. In some of these systems, especially membranes, the flow relation of solutions are complicated by interactions between the solvent (water) and solutes. These will be discussed in Chapter 11; for the present, it will be assumed, when dealing with water, that flow is unaffected by such interactions. This is a reasonably safe assumption in the case of soils and certain artificial membranes.

Darcy's law

The classical work in this field was carried out over a century ago by Darcy who studied the seepage rates of water through filter beds under different hydraulic pressure heads (Fig. 3.1). He showed that the volume rate of flow q_v through the bed was directly proportional to the pressure head (here represented by the height of a water column Δh) and the cross sectional area of the bed (A), and inversely proportional to the thickness of the bed (l).

Darcy's equation (or law) may be represented in the form:

$$q_v = KA \frac{\Delta h}{l}. \tag{3.1}$$

The proportionality constant K is a measure of the permeability of the medium to water, or as it is usually known, its hydraulic conductivity. With q_v expressed as a volume per unit time ($L^3\,T^{-1}$), A as an area (L^2) and l as a length (L), K has the dimensions of a velocity (LT^{-1}) and is expressed in units of m s^{-1}.

If we replace q_v by ϕ_v, the flux of water through the medium (i.e. volume rate of flow per unit area) and Δh by Δp, the corresponding pressure difference ($\Delta p = \rho g \Delta h$), we can express Darcy's law in a more general form:

$$\phi_v = \frac{q_v}{A} = \frac{K\,\Delta p}{\rho g l} = K' \frac{\Delta p}{l}. \tag{3.2}$$

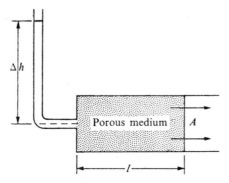
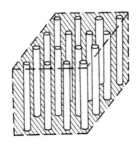

FIG. 3.1. Simplified model illustrating Darcy's   FIG. 3.2. Model of porous medium
law (for explanation see text p. 29).   with uniform, circular, parallel pores.

This illustrates rather more clearly the underlying principle, namely that the flux ϕ_v is proportional to the pressure gradient $\Delta p/l$ in the direction of flow. If Δp is expressed as N m^{-2} (dimensions ML^{-1} T^{-2}) then the new hydraulic conductivity coefficient, K' will have the dimensions M^{-1} L^3 T, with units of m^3 kg^{-1} s.

Even in this form however, Darcy's law is still of limited application because the coefficient K' reflects not only the conducting properties of the medium, but also the viscosity of the fluid, which itself influences flow. To separate these factors another coefficient, k is introduced, defined by:

$$k = K'\eta \qquad \text{where } \eta \text{ is the dynamic viscosity.} \qquad (3.3)$$

Equation (3.2) can now be written:

$$\phi_v = \frac{q_v}{A} = \frac{k\,\Delta p}{\eta l}. \qquad (3.4)$$

The coefficient k is known as the *intrinsic* or *specific permeability* (sometimes simply the permeability) and now refers exclusively to the medium. Inspection of the dimensions of eqn (3.4) shows that k has the dimensions L^2 and units of m^2.

Much confusion has arisen in the literature because of the failure of many authors to distinguish clearly between K, K', and k.

Compressible gases

So far it has been implied that Darcy's law, as expressed above, applies to all fluids on the assumption that they are incompressible. This is probably true under most circumstances, even with gas flow, but there are occasions, as in certain measurements in wood science (p. 37) when gases are forced through a porous medium under pressure, and account has then to be taken of changes in volume as the pressure decreases along the line of flow. Darcy's law must then be expressed in the differential form:

$$\phi_v = -\frac{k}{\eta} \cdot \frac{dp}{dl} \qquad (3.5)$$

(negative because p decreases with increasing l).

As shown in Appendix 4 the solution to this is:

$$\phi_v = \frac{k}{\eta} \cdot \frac{\bar{p}}{p} \cdot \frac{\Delta p}{l}, \qquad (3.6)$$

where p corresponds to the applied pressure and $\bar{p}$ to the average pressure along the length l of medium.

If the pressure drop Δp is so small that $\bar{p} = p$, compressibility can be ignored and eqn (3.6) then reverts to eqn (3.4) for incompressible fluids.

Permeability and pore geometry

Numerous attempts have been made to relate quantitatively, the permeability of a porous medium to certain of its geometrical properties. For example, if we assume the simplest geometry and treat the medium as a parallel array of uniform circular tubes embedded in a non-conducting matrix (Fig. 3.2) then if flow is laminar, the permeability k can readily be derived from Poiseuille's law.

If a given cross-sectional area A of medium contains N pores of radius r, the total pore area A_p is $N\pi r^2$. If the pressure gradient $= \Delta p/l$, then from Poiseuille's law, the total volume rate of flow across the pore area will be:

$$q_v = \frac{N\pi r^4 \Delta p}{8\eta l} = \frac{A_p r^2 \Delta p}{8\eta l}. \qquad (3.7)$$

Substituting for q_v from eqn (3.4) gives:

$$k = \frac{A_p}{A} \times \frac{r^2}{8}. \qquad (3.8)$$

In other words the permeability is directly proportional to the product of the pore fraction A_p/A and the square of the radius r^2.

This simple relationship has been used to calculate the pore radius of artificial membranes from the pore fraction (as determined from the wet and dry weights of the membrane) and the permeability k (as determined by direct measurement) (Pappenheimer 1953). Strictly speaking, eqn (3.8) only applies to media containing cylindrical pores of uniform size. If the sizes vary, the flow rate through the larger pores will be higher than that through the smaller pores according to the 4th power of the ratio of their radii. However, an effective radius, $\bar{r}$ can be defined such that:

$$\bar{r} = \sqrt[4]{(f_1 r_1^4 + f_2 r_2^4 + f_3 r_3^4 \ldots f_n r_n^4)} \tag{3.9}$$

where r_1, r_2, r_3 etc. represent different size classes of pores and f_1, f_2, f_3 etc. their respective frequencies, i.e. fractions of the total number of the pores present. It will be appreciated that the effective pore radius, calculated from eqn (3.9) will always be greater than the arithmetic mean of the pore size distribution.

That reasonably good agreement has been obtained between calculated values of $\bar{r}$ and values obtained by other, independent methods, suggests that for certain membranes at least, the simple model illustrated in Fig. 3.2 applies. With most porous media however, apart from variations in pore size, it cannot be assumed that the pores are cylindrical or that they are straight and continuous through the medium. Several models have been proposed which attempt to take account of these complications.

One of the best known of these models is the Kozeny–Carman equation:

$$\phi_v = \frac{q_v}{A} = \frac{\epsilon^3}{cs^2} \times \frac{\Delta p}{\eta l}, \tag{3.10}$$

where ϵ corresponds to the porosity of the medium, i.e. the total pore volume per unit volume of medium and s to the total internal surface area of the pores per unit volume of medium. As before, Δp represents the pressure drop across a thickness l of medium in the direction of flow and A the cross-sectional area. The factor c, known as the Kozeny–Carman constant, makes allowance for the shape of the pores and the tortuosity of the pathway through the medium (see Appendix 5).

Since Darcy's equation (eqn 3.4) states that $\dfrac{q_v}{A} = \dfrac{k\Delta p}{\eta l}$ the intrinsic permeability k for the Kozeny–Carman equation is given by:

$$k = \frac{\epsilon^3}{cs^2}. \tag{3.11}$$

Some authors (e.g. Hillel 1971) quote eqn (3.11) in a rather different form using instead of s, a so-called specific surface factor (a or s_0) defined as the total internal surface area per unit volume of *solid*. This has certain advantages over s since s_0 is independent of the porosity whilst s is not. The two factors are related by the equation, $s = s_0(1-\epsilon)$.

The Kozeny–Carman equation has been widely used for determining the saturation permeability of unconsolidated porous media, i.e. media made up of discrete particles, such as spheres. It has proved particularly useful when the pores are of fairly uniform size, though much depends on how accurately the internal surface area can be measured. For most practical purposes,

$c = 5$ covers a wide range of structures with reasonable accuracy. However, the equation is not suited to media with a wide range of pore sizes such as soil.

Soil pores are typically irregular, tortuous, and interconnected, sometimes ending blindly. Depending on the soil type, e.g. sand or clay, they may vary from large voids connected by narrow necks to a more complicated pattern induced by aggregation, as found in crumb structures. Furthermore, account must be taken of the variable moisture content. When saturated, almost the whole of the soil pore space is available for flow and the permeability will then be a maximum, though in clays, the situation is then complicated by swelling. When the soil moisture falls, as through abstraction by plant roots, the largest pores are the first to empty and flow becomes more and more restricted to the smaller pores with a resulting fall in conductivity (see Fig. 11.4). The most rewarding approach to the study of soil permeability under these conditions is undoubtedly that based on soil pore-size distribution and interaction and one of the most successful models using this approach is that of Millington and Quirk (1960).

The basic argument is that flow across a plane section of the soil can only be maintained if the pores on either side are continuous, and that the total effective pore area for flow produced by a more or less random association of pores across such a plane (the pore interaction) approximates to $\epsilon^{4/3}$ where ϵ is the porosity of the soil. Since the permeability is determined by the effective pore area and the square of the pore radius (eqn 3.8), the pores are divided into several size classes according to their radius (r_1, r_2 etc.) and corresponding areas (a_1, a_2 etc.). By matching pairs of pores in all possible combinations ($a_1 \times a_2$, $a_1 \times a_3$, $a_2 \times a_3$ etc.) a series of effective areas is obtained; the product of each of these areas and the square of the radius of the smaller pore of the pair (assuming that the smaller pore will control the flow) then gives a measure of the contribution of each pore combination to the overall permeability. The latter is obtained by summating these contributions. If each size class occupies the same proportion of the total porosity and there are m size classes, the final expression for permeability becomes:

$$k = \frac{1}{8} \frac{\epsilon^{\frac{4}{3}}}{m^2} [r_1^2 + 3r_2^2 + 5r_3^2 + \ldots(2m-1)r_m^2] \tag{3.12}$$

The number of size classes is usually about 10.

For unsaturated flow, when only some of the pores are filled with liquid and are effective in conduction, the value of ϵ is taken as that of the liquid-filled pore space and the r^2 series begins with the largest size-class containing liquid. The relation between pore size and moisture content is determined by experiment.

A further advantage of this model is that it can also be applied, though with certain reservations, to consolidated porous media in which the solid forms a continuous and permanent structure enclosing the pores, such as sandstones and limestones. Because of difficulties in determining the internal surface area of the pores in these consolidated media, the Kozeny–Carman equation cannot easily be applied.

Fluid movement in wood

A considerable number of studies have been made on the flow of water, other liquids, and gases in wood, partly stimulated by interest in the transport of sap in plants, especially in trees, and partly because of technical problems such as those associated with the drying of wood and with its impregnation by preservatives.

In conifers, the conducting elements of the xylem comprise the tracheids, elongated tapering cells, about 1 to 7·5 mm long and 15 to 80 μm in diameter which communicate with each other through small pits to form a continuous conducting system (Fig. 3.3). The pits are of the 'bordered' type, characterized by an impermeable disc or torus at the centre which acts as a kind of valve controlled by pressure differences on either side. When the 'valve' is open (the so-called unaspirated state), fluids can pass from tracheid to tracheid via small pores, of an equivalent diameter of the order of 0·1 to 4 μm, in the membrane surrounding the torus (inset, Fig. 3.3). These pores impose a relatively high resistance to both longitudinal and lateral flow.

The angiosperm conducting system is very much more complex and variable and only the barest outline can be given here. Longitudinal flow occurs almost exclusively in the lumina of the vessel elements, which range in diameter from about 15 to 150 μm in diffuse-porous species and from about 60 to 400 μm in ring-porous species (Fig. 3.4). These two groups are distinguished mainly by the size distribution of the vessels within the annual ring (Fig. 3.5). The vessel elements are joined end to end, and with the virtual disappearance of most of the cross walls, they provide uninterrupted pathways extending for about 0·8 to 1·5 m in diffuse-porous, and 0·5 to 18 m in ring-porous species. Lateral flow is possible, though against a much greater resistance, via pits between contiguous vessels and between vessels and other woody elements.

Besides specific differences in wood structure, there is often an appreciable variation within species in the size, frequency, and distribution of the conducting elements. In older stems, a distinction must also be made between the outer conducting sapwood and the inner heartwood, which for various reasons, no longer acts as a conducting tissue.

On the basis of Darcy's law, numerous measurements have been made of the hydraulic conductivity K' of stems and timbers by measuring the volume flow of water through a specimen of known dimensions under a given pressure gradient (eqn 3.2). By taking into account the viscosity of water, the corresponding values for the intrinsic permeability k may be calculated (eqn 3.3). Since wood swells with water, more precise measurements can be made using non-swelling liquids such as n-hexane.

Some of the earliest and most comprehensive measurements of conductivity were made by Farmer (1918) on stems of a variety of woody species: his results, summarized by Heine (1971) and converted into SI units, indicate k values ranging from 3×10^{-13} to 4×10^{-12} m² with conifers and evergreens giving the lowest values and ring-porous trees the highest.

In living stems only a fraction of the conducting elements appear to be functional. Because of heartwood formation, conduction in the larger stems of conifers and diffuse-porous species is usually confined to the few outermost annual rings constituting the sapwood, whilst in most ring-porous trees, only the vessels of the single outermost ring transport sap.

If lateral flow is neglected, hardwoods can be considered as approximating to the porous model of Fig. 3.2, but with a greater or lesser variation in pore size. By means of eqn (3.8), an estimate can be made of the fraction of the xylem cross-sectional area occupied by functional vessels from a knowledge of k and the vessel dimensions. Since this equation involves the square of the radius, variations in size can be allowed for by taking the root mean square. For example, according to Farmer, k for beech stems ranges from about 15 to 38×10^{-13} m², with a mean value of 27×10^{-13}, and according to Zimmerman (1971) the vessel radii in this species range from 8 to 40 μm; assuming a uniform distribution of vessel size (Fig. 3.5b), the root mean square will be about 30 μm. Substituting $k = 27 \times 10^{-13}$ and $r = 30 \times 10^{-6}$ in eqn (3.8) gives $A_p/A = 0.024$ i.e., the area of conducting lumina is about 2.4 per cent of the total area of xylem tissue, or about 12 per cent of the total vessel area assuming the latter to occupy 20 per cent of the area of xylem tissue. This value agrees closely with those quoted by Heine (1970) for beech and other diffuse porous species.

Another approach to the problem of sap conduction in tree stems is to measure the velocity of flow. This has been done on standing trees using a variety of techniques including dyes, radioactive tracers, and heat convection (Leyton 1971). Published maximum rates range from about 1 to 2 m h⁻¹ in conifers, from 1 to 6 m h⁻¹ in diffuse-porous trees and from 4 to 40 m h⁻¹ in ring-porous trees (Zimmerman 1971). Given the vessel dimensions, eqn 2.4, should allow us to estimate the pressure gradients required to propel the sap at these velocities, assuming that they correspond to mean values. For example, the highest recorded velocity is 43.6 m h⁻¹ (1.2×10^{-2} m s⁻¹) for an oak tree in which the larger vessels have radii ranging from 100 to 150 μm (Fig. 3.5a). Assuming a root mean square radius of 130 μm, and taking the viscosity of sap as approximately the same as for water (10^{-3} kg m⁻¹ s⁻¹), eqn (2.4) predicts a pressure gradient ($\Delta p/l$) of 5.7×10^3 N m⁻² per m (about 0.06 bar per m).

The above estimate is almost certainly too low. Except for one or two species, pressure gradients calculated from eqn (2.4) fall well below those found by measurement on detached stems, in certain species by as much as

4

eight times (Zimmerman 1971). The main source of the discrepancy are the cross walls which interrupt flow at intervals along the stem. With ring-porous species the discrepancy is probably not so large, though even with oak, the pressure gradient would be at least three times that calculated from eqn (2.4), i.e. about 20×10^3 N m^{-2} per m. In standing trees we must add to this an extra pressure gradient of approximately 1×10^4 N m^{-2} per m to allow for the hydrostatic gradient, giving a total gradient of about 3×10^4 N m^{-2} (approximately 0·3 bar) per m.

That the above estimate is of the right order of magnitude is supported by a rather different approach suggested by Heine (1970, 1971). He proposes a so-called conductivity coefficient, σ (units, m^4 N^{-1} s^{-1}) defined by:

$$u_{\mathrm{m}} = \sigma \frac{\Delta p}{l},\tag{3.13}$$

where $u_{\mathrm{m}} =$ average velocity of flow.

The coefficient σ may be determined directly from velocity measurements made on detached stems or may be calculated from k as follows. The volume flow rate q_{v} across an area A of porous medium containing a functional lumen area A_p is given by $q_{\mathrm{v}} = u_{\mathrm{m}} A_p$. Substituting for q_{v} in eqn (3.4) gives:

$$u_{\mathrm{m}} = \frac{Ak \, \Delta p}{A_p \eta l}.$$

Hence by analogy with eqn (3.13):

$$\sigma = \frac{Ak}{A_p \eta} \quad \text{and} \quad \frac{\Delta p}{l} = \frac{u_{\mathrm{m}}}{\sigma} = \frac{u_{\mathrm{m}} A_p \eta}{Ak}.\tag{3.14}$$

For oak, Farmer quotes a maximum value equivalent to about

$$k = 100 \times 10^{-13} \text{ m}^2$$

and measurements on this species suggest that the vessels occupy about 7 per cent of the total xylem area. Assuming that 20 per cent of these vessels are functional gives $A_p/A = 0.014$. Substituting this value in eqn (3.14) with $\eta = 10^{-3}$ kg m^{-1} s^{-1} and, as before, $u_{\mathrm{m}} = 1.2 \times 10^{-2}$ m s^{-1} gives $\Delta p/l = 17 \times 10^3$ N m^{-2} per m. It will be appreciated that in the absence of precise data for the material, the agreement between this and the previously calculated gradient may be fortuituous. However, on the assumption that a total gradient of about 3×10^4 N m^{-2} per m (including the hydrostatic gradient) is of the right order, the total driving pressure for a tree 50 m tall would be of the order of 15×10^5 N m^{-2} (about 15 bar). It is now generally accepted that the driving force responsible for the ascent of sap in trees derives from the tension developed in the leaves as a result of evaporation of water (transpiration) from the walls of the mesophyll cells (see p. 185). Measurements have confirmed that tensions of the order of 20×10^5 N m^{-2} (about

20 bar) can readily be developed in the foliage of actively transpiring trees, thus more than enough to 'pull' the sap up stems at the maximum recorded velocity in tall oak trees. The problem of sap movement in plants will be discussed in more detail in Chapter 11.

Although certain authors have been tempted to apply the above calculations to the tallest trees (Sequoia), which grow up to heights of 100 m or so, we have as yet no precise data on sap velocities in these species. In any case, the Sequoias are conifers for which the simple model of Fig. 3.2 cannot strictly be used because of the tortuous pathway followed by the sap from tracheid to tracheid through the interconnecting pores.

Various models for fluid transport in conifer wood have been proposed (Siau 1971) and a variety of techniques, including gas permeability measurements, have been applied to determine pore size distributions and numbers in the pit membranes. On the assumption that the resistance of the tracheid lumina can be ignored, the common approach in gas permeability measurements has been to apply the equation, attributed to Adzumi:

$$\frac{q_v p}{\Delta p} = \frac{N \pi r^4 \bar{p}}{8 \eta l} + \frac{4}{3} \frac{N r^3}{l} \sqrt{\left(\frac{2 \pi R T}{M}\right)}, \tag{3.15}$$

where q_v = volume flow rate measured at pressure p with a pressure difference Δp across the ends of the specimen and with an average pressure $\bar{p}$. The first quantity on the right hand side of eqn (3.15) refers to viscous flow through a medium containing N pores of radius r and length l in the area across which q_v is measured; this can readily be derived from the differential form of Poiseuille's equation along the lines indicated in deriving eqn (3.6). The second quantity makes allowance for some slip flow through the pores (p. 28), though strictly speaking, this only applies to pores so small that slip flow predominates. Plotting $q_v p/\Delta p$ against $\bar{p}$ should then give a straight line from which r and N can be obtained from the slope and intercept. According to Petty (1970) however, the relation for spruce wood is not linear because the tracheid lumina make a significant contribution (about 9 per cent of the total) to the resistance to gas flow and therefore could not be ignored. His experimental curves could be resolved into two linear components, one relating to the resistance of the tracheid lumina, the other to that of the membrane pores; from the latter, modified to allow for the small depth of the pores, he was then able to calculate their equivalent radius and number.

4. Flow over plane surfaces

WE have already seen (p. 11 and Fig. 1.7) that when a fluid moves over a plane surface (or when the surface moves relative to the surrounding fluid), a laminar boundary layer is formed over the front part of the surface and then, if the critical Reynolds' number is exceeded, flow becomes turbulent. As in the case of pipe flow, we shall begin by deriving certain quantitative relationships for both types of flow and then discuss their biological implications. Because of its special importance as an environmental factor, wind flow over the ground will be discussed separately in Chapter 9.

Laminar flow

The curve OA in Fig. 4.1 depicts the growth of the laminar boundary layer over a flat plate with increasing distance from O, the leading edge. Also shown is the velocity profile within this layer at C, distance l from O.

Theoretically, since the fluid in the boundary layer merges imperceptibly into the free stream above, the depth of this boundary layer cannot be defined precisely. In fact, for an exact solution, it is necessary to assume that the velocity of flow within the boundary layer approaches that of the free stream asymptotically. However, it is convenient in practice to assume a definite depth δ_L at the limit of which the velocity u equals that of the free stream U. Assuming further that the shear stress decreases linearly with increasing distance z normal to the surface, it can be shown (Appendix 6) that the depth of the laminar boundary layer approximates to:

$$\delta_L \approx 5 \sqrt{\left(\frac{\eta l}{\rho U}\right)} \approx 5l(Re_l)^{-\frac{1}{2}}, \tag{4.1}$$

where l refers to the distance along the plate from the leading edge, and (Re_l) to the corresponding Reynolds' number (Ul/ν).

We see from eqn (4.1) that the depth of the laminar boundary layer increases directly as the square root of the distance from the leading edge ($l^{\frac{1}{2}}$) and inversely as the square root of the free stream velocity ($U^{-\frac{1}{2}}$). When the free stream flow is turbulent, U will represent the velocity averaged over a period of time. The numerical factor 5 must be treated as a working approximation; depending on the assumptions made in solving the problem, other

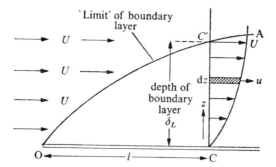

FIG. 4.1. The growth of the laminar boundary layer over a flat plate with velocity profile indicated at C, distance l from the leading edge O. Free stream velocity $= U$. For details see Appendix 6.

slightly different values have been suggested, but the rest of the equation remains unaffected.

Within the boundary layer, the shape of the velocity profile is parabolic (Appendix 6). Unlike the situation in pipes however, the drag on a plate due to skin friction τ_0 will vary with the distance along the plate and correspondingly, with the depth of the boundary layer. At a distance l from the leading edge the theoretical solution gives:

$$\tau_0 = 0 \cdot 332 \rho U^2 (Re_l)^{-\frac{1}{2}}, \tag{4.2}$$

where as before, (Re_l) corresponds to the Reynolds' number for the distance l. In order to evaluate the total drag force F_{DL} over the whole area of the plate occupied by the laminar boundary layer, τ_0 given in eqn (4.2) must be integrated from O to l, corresponding to the area of a rectangular plate of unit width; this works out as:

$$F_{DL} = 0 \cdot 664 \rho U^2 l (Re_l)^{-\frac{1}{2}}. \tag{4.3}$$

Since $(Re_l) = Ul/\nu$, this means that the drag exerted by a moving fluid on a plane surface increases as the square root of its length $(l^{\frac{1}{2}})$ in the direction of flow and as the 3/2 power of the free stream velocity $(U^{\frac{3}{2}})$. If the plate is completely immersed in the fluid so that both surfaces are involved, the drag force will be double that given above, and for a rectangular plate of width b, the value for F_{DL} must be further multiplied by b.

Turbulent flow

No complete theoretical solution is available for turbulent flow, but a working solution can be obtained from measurements on turbulent flow in pipes. As shown in Appendix 6, the depth of the turbulent boundary layer

(δ_T) over a flat plate at a distance (l_T) from the *beginning of turbulence* (*not* the leading edge) is given by:

$$\delta_T = 0\cdot376 l_T (Re_{tT})^{-\frac{1}{5}}, \tag{4.4}$$

where (Re_{tT}) refers to the Reynolds' number based on (Ul_T/ν).

It follows that the turbulent boundary layer increases in depth as $l_T^{\frac{1}{5}}$ i.e. more rapidly than the laminar boundary layer which increases as $l^{\frac{1}{2}}$.

It can also be shown that the local skin friction (τ_0) is related to δ_T, the depth of the turbulent boundary layer by:

$$\tau_0 = 0\cdot023\rho U^2 \left(\frac{\nu}{U\delta_T}\right)^{\frac{1}{4}}. \tag{4.5}$$

Substituting for δ_T from eqn (4.4) gives:

$$\tau_0 = 0\cdot029\rho U^2 (Re_{tT})^{-\frac{1}{5}}. \tag{4.6}$$

The total drag force F_{DT} exerted by the fluid on one side of the plate is then obtained by integrating τ_0 over the area of the plate occupied by the turbulent boundary layer, and assuming a rectangular plate of unit width, this works out as:
$$F_{DT} = 0\cdot0366\rho U^2 l_T (Re_{tT})^{-\frac{1}{5}} \tag{4.7}$$

Thus, whereas with laminar flow the total drag force is proportional to $l^{\frac{1}{2}}$ and $U^{\frac{3}{2}}$, with turbulent flow it is proportional to $l^{\frac{4}{5}}$ and $U^{\frac{9}{5}}$; in other words the drag increases far more rapidly with increasing surface length when flow becomes turbulent. As we shall see later, this is of some importance in the energy relations of swimming and flying organisms large enough to produce a turbulent boundary layer.

As before, F_{DT} refers to a single surface of unit width; it must therefore be multiplied by two if the plate is completely immersed so that both surfaces are involved, and further by b for a plate of width b, assuming that it is rectangular.

It will be appreciated that in practice, if the surface is long enough, and/or the fluid velocity high enough, both a laminar and turbulent boundary layer will be formed. In this case two calculations will normally have to be made to calculate the drag, the first for the length of plate l occupied by the laminar boundary layer, the second for the remaining length of plate l_T occupied by the turbulent boundary layer. The length (more strictly the zone) at which transition occurs will be determined by the critical Reynolds' number, which varies with the level of turbulence in the approaching free stream and also with the roughness of the surface. When the free stream turbulence is low and the surface is smooth, transition may not occur until a very high (Re) is reached, possibly as high as 1×10^6 to 3×10^6. In practice however, as in the case of wind or water flow in streams, some turbulence is almost always present and if it is not too great, a useful working value for most surfaces is

5×10^5. Increasing surface roughness will tend to promote the onset of turbulence.

The laminar sub-layer

The laminar sub-layer which occurs next to the surface may be of some biological importance partly because in the transport of heat and matter between a surface and the surrounding medium, it can only support molecular diffusion, and partly because, as we have already seen (p. 25), its depth determines the extent to which roughness elements on the surface affect the drag. The depth of this layer (δ_s) may be calculated along the lines described for turbulent flow in pipes (Appendix 3). Assuming the same working value of $\delta_s = 10\nu/u_* = 10\nu/\sqrt{(\tau_0/\rho)}$, and substituting for τ_0/ρ from eqn (4.6), we obtain:

$$\delta_s \approx \frac{58l_T}{(Re_{l\mathrm{T}})^{0 \cdot 9}} \quad \text{or, very approximately,} \quad \delta_s \approx \frac{58\nu}{U}. \quad (4.8)$$

In air therefore, with a speed $U = 5 \text{ m s}^{-1}$ (about 10 miles h^{-1}) and $\nu = 1 \cdot 5 \times 10^{-5} \text{ m}^2 \text{ s}^{-1}$, the depth will be about 0·17 mm. It will decrease still further as the windspeed U increases.

Friction coefficients

As in the case of pipe flow, it is convenient to express the frictional drag on a plate in terms of a drag coefficient by dividing the drag force per unit area of surface by $\frac{1}{2}\rho U^2$. In the case of a plate however (and as we shall see later, also for other surfaces), we must distinguish between a local skin friction force (τ_0) which varies with the distance from the leading edge, and the average skin friction for the whole of the surface concerned. The latter will equal the total drag force F_D divided by the area of the surface; for a plate of unit width, this means dividing F_D by the length of plate involved.

We can therefore define two friction coefficients, a local coefficient C_t and an average coefficient C_F:

$$\text{for laminar flow (eqn 4.2), the local } C_t = 0 \cdot 664 \ (Re_l)^{-\frac{1}{2}} \quad (4.9)$$

$$\text{for turbulent flow (eqn 4.6), the local } C_t = 0 \cdot 058 \ (Re_{l\mathrm{T}})^{-\frac{1}{5}} \quad (4.10)$$

(n.b. τ_0 is already expressed on a unit area basis);

$$\text{for laminar flow (eqn 4.3), the average } C_F = 1 \cdot 328 \ (Re_l)^{-\frac{1}{2}}; \quad (4.11)$$

$$\text{for turbulent flow (eqn 4.7), the average } C_F = 0 \cdot 073 \ (Re_{l\mathrm{T}})^{-\frac{1}{5}}. \quad (4.12)$$

The change in C_F with increasing (Re) for both laminar and turbulent flow is illustrated in Fig. 4.2, and as in the case of the pipe friction coefficient (Fig. 2.6), shows the marked increase in drag when turbulent flow prevails.

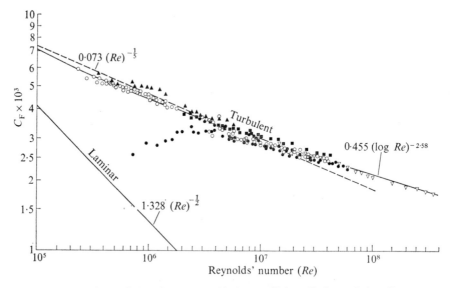

FIG. 4.2. Experimental data for average friction coefficient C_F for turbulent flow over a smooth flat surface, plotted as a function of the Reynolds' number (*Re*). The theoretical relationship for laminar flow is included. Redrawn from data of Rohsenow and Choi (1961).

We shall discuss the biological significance of this later. Note also that at (*Re*) > 10^7, a rather better fit to the experimental data is given by

$$C_F = 0.455(\log Re)^{-2.58}.$$

Drag on rough surfaces

As with flow in pipes, the roughness of the surface has no effect on the drag when laminar flow prevails, but depending on the height of the roughness elements in relation to the depth of the laminar sub-layer, roughness may lead to a substantial increase in the drag with turbulent flow. As we shall see later this may be of some biological importance and it becomes of special significance in the case of wind flow over rough ground (Chapter 9).

The approach we shall adopt here is similar to that for rough pipes. If k represents the average height of the roughness elements on the surface, the local C_t can be expressed empirically by:

$$C_t = \left[2.87 + 1.58 \log_{10} \frac{l_T}{k} \right]^{-2.5}, \tag{4.13a}$$

where l_T refers to the distance from the start of the turbulent boundary layer. The average coefficient C_F is then given by:

$$C_F = [1.89 + 1.62 \log_{10} l/k]^{-2.5}. \tag{4.13b}$$

Biological implications

Assuming that the critical (Re) of 5×10^5 represents conditions appropriate to surfaces exposed to wind or water, we can readily calculate the length of surface l required to induce turbulence from $Ul/\nu = 5 \times 10^5$, whence $l = 5 \times 10^5 \, \nu/U$. For air at 20°C, $\nu = 1.51 \times 10^{-5}$ m² s⁻¹, therefore, $l = 7.6/U$ (m), and for water, with $\nu = 10^{-6}$ m² s⁻¹, $l = 0.5/U$ (m) where U = velocity of the free stream in m s⁻¹.

Thus, for a wind-speed of 0·447 m s⁻¹ ($= 1$ mile per h), l would need to exceed 17 m, whilst at 8·9 m s⁻¹ (≈ 20 miles per h), it would still need to be greater than about 0·8 m for turbulence to occur. Except at very high wind speeds therefore, only a laminar boundary layer would be expected over the surfaces of bodies (e.g. leaves or the flat surfaces of animal bodies) of normal dimensions. The situation may be rather different in an aqueous medium; in this case, for the same velocities, the length need only be about $\frac{1}{15}$th of that in air for turbulence to be initiated (since $\nu_{\text{water}} \approx \frac{1}{15}\nu_{\text{air}}$). Turbulent boundary layers would therefore be expected with large, fast swimming marine organisms. It must be remembered of course, that in all cases, a length l of surface will always be occupied by a laminar boundary layer, though at high (Re), as in the case of the dolphin (p. 55), it may be negligible.

It is of some interest to compare the drag on surfaces due to laminar and turbulent flow in air and water. For laminar flow we can calculate the drag from eqn (4.3) assuming a critical (Re) of 5×10^5; as shown above, if the surface is long enough, a length of $7.6/U$ (m) for air and $0.5/U$ (m) for water will be occupied by the laminar boundary layer. For air, therefore, taking $\rho = 1.2$ kg m⁻³, the total drag (per unit width) will be

$$F_{\text{DL}} = 0.664 \times 1.2 \times U^2 \times 7.6/U(5 \times 10^5)^{\frac{1}{2}} = 0.0086U \quad \text{(N)}.$$

With a wind speed of 8·9 m s⁻¹, $F_{\text{DL}} = 0.076$N.

For water at 20°C, $\rho = 10^3$ kg m⁻³ and $F_{\text{DL}} = 0.47U$ N; with the same velocity of 8·9 m s⁻¹, $F_{\text{DL}} = 4.18$ N (per unit width) i.e. about 55 times that for air.

To evaluate the drag due to turbulent flow, we need to know the length of plate involved, i.e. l_{T}. For a critical (Re) of 5×10^5, the above calculations indicate that l, the length of surface occupied by the laminar boundary layer, will be 0·85 m for air and 0·06 m for water, assuming $U = 8.9$ m s⁻¹. If the surface was 1 m long in the direction of flow, the rest would be occupied by a turbulent layer (ignoring the transition zone); l_{T} therefore, amounts to $1 - 0.85 = 0.15$ m in air, and $1 - 0.06 = 0.94$ m in water. For these lengths, $(Re_{l_{\text{T}}}) = Ul_{\text{T}}/\nu = 8.8 \times 10^4$ for air and 8.4×10^6 for water. We can now calculate the drag force F_{DT} from eqn (4.7). Thus, for air,

$$F_{\text{DT}} = 0.0366 \times 1.2 \times (8.9)^2 \times 0.15 \times (8.8 \times 10^4)^{-\frac{1}{5}} = 0.054 \text{ N (per unit width)}.$$

The total drag, due to both laminar flow (F_{DL}) and turbulent flow (F_{DT}) will then be $0.076 + 0.054 = 0.13$ N (per unit width).

The same calculation for water gives $F_{DT} = 112$ N and $F_{DL} + F_{DT} = 116$ N per unit width. In other words, under the stated conditions, the drag induced by water as a medium is almost 900 times that in air. Most of this enormous difference between the two media must be attributed to the very much higher density of water, since the greater length of plate occupied by the turbulent boundary layer in water is offset to some extent by a higher (Re).

As already explained, the nature of boundary layer flow over a surface has important consequences in relation to the transport of heat and other properties between the surface and the surrounding medium; these will be discussed in later chapters. At this stage we might turn to flow over curved surfaces which are much more common in nature than plane surfaces.

5. Flow over immersed bodies

THIS Chapter is concerned mainly with the nature and extent of the drag forces which resist the movement of a body in a fluid. Clearly, this problem is particularly relevant to the energy relations of swimming and flying organisms. We have already seen (p. 11) that when a fluid moves over a plate lying with its plane in the line of flow, its movement is resisted by viscous frictional forces within the boundary layer; conversely, if the plate was moved with the same speed relative to the fluid, it would experience the same frictional drag. Almost all living organisms however, possess curved surfaces, their bodies commonly corresponding to spheres, cylinders or in many cases to streamline shapes. Fluid behaviour over such bodies is much more complex than over plane surfaces. When travelling through water or air, all will suffer some frictional drag, but, depending on their shape, size and speed, they will generally experience an extra drag known as *form* or *pressure drag*. In contrast to friction drag, pressure drag originates in the bulk fluid outside the boundary layer. However, before we can deal with this we must first consider certain underlying principles, among the most important of which is that due to Bernoulli.

The Bernoulli equation

This equation describes a very important relationship between the velocity of a fluid and its pressure. As an example let us consider what happens when a fluid flows through a circular pipe which decreases in diameter over a relatively short length (Fig. 5.1).

If, by means of manometers, we were to measure the pressure within the fluid, we would find that the pressure in the wider tube A exceeded that in the narrower tube B by an amount depending on the mean velocities of flow in the two tubes (u_{m1}, u_{m2} respectively). For a given volume rate of flow, these mean velocities will of course be inversely proportional to the cross-sectional areas of the tubes.

The precise relationship between pressure and velocity is derived from a consideration of the mechanical energy of a non-viscous incompressible fluid in motion. Picture a unit volume of fluid moving from A to B in the system illustrated in Fig. 5.1. If the density of the fluid is ρ, the mass of this

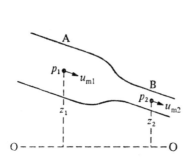

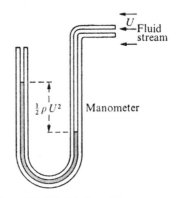

FIG. 5.1. Diagram illustrating Bernoulli's theorem. For explanation see text p. 45.

FIG. 5.2. Pitot tube with manometer for measuring fluid velocities.

unit volume will also be ρ. Let us suppose further that the tube slopes so that the two points at which the pressures and velocities are measured are at heights z_1, z_2 respectively, above some arbitrary datum level O–O.

At A, the fluid element will possess, in relation to O–O, a potential energy equal to $\rho g z_1$, where g is the acceleration due to gravity. Because it is in motion, it will also possess a certain kinetic energy equal to $\frac{1}{2}\rho u_{m1}^2$. Finally if it exerts a pressure p_1, it is capable of doing the equivalent amount of mechanical work. The total mechanical energy E_A of our unit volume of fluid at A is then given by:

$$E_A = \rho g z_1 + \tfrac{1}{2}\rho u_{m1}^2 + p_1$$

By the same argument, the total mechanical energy E_B of the same unit volume of fluid at B, where the pressure is p_2 and the velocity u_{m2}, will be given by:

$$E_B = \rho g z_2 + \tfrac{1}{2}\rho u_{m2}^2 + p_2$$

As the fluid is assumed non-viscous, no energy will be lost in overcoming skin friction between A and B, and if there is no other expenditure or income of energy, $E_A = E_B$, whence:

$$\rho g z_1 + \tfrac{1}{2}\rho u_{m1}^2 + p_1 = \rho g z_2 + \tfrac{1}{2}\rho u_{m2}^2 + p_2.$$

This is Bernoulli's equation and dividing throughout by ρg gives the form in which it is usually expressed:

$$z_1 + \frac{1}{2}\frac{u_{m1}^2}{g} + \frac{p_1}{\rho g} = z_2 + \frac{1}{2}\frac{u_{m2}^2}{g} + \frac{p_2}{\rho g} \tag{5.1}$$

An important consequence of this equation is that if there is no change in $z_1 - z_2$, any increase in fluid velocity must be accompanied by a decrease in its pressure, and vice versa. Although in theory, this relationship applies strictly to non-viscous fluids, the inverse relationship between velocity and pressure still holds for viscous fluids such as water and air.

A simple everyday example of the Bernoulli principle is the common scent or paint spray in which air under pressure is blown over the end of a vertical tube, the bottom end of which dips into the liquid. As a result of the high velocity of the air issuing from the nozzle, its pressure is reduced and if it falls sufficiently below atmospheric pressure, the latter will force the liquid up the vertical tube from which it is 'atomized' to produce a spray. Much the same principle underlies the operation of the common laboratory filter pump and the various devices used for measuring fluid flow rates such as the Pitot tube, Venturi meter, etc.; several of these latter devices involve the measurement of a pressure difference resulting from a change in velocity, usually induced as in Fig. 5.1, by changing the pipe dimensions. The Pitot tube consists essentially of a narrow bore tube with one end facing directly into the oncoming fluid stream and the other connected to a manometer or other device for recording pressure differences (Fig. 5.2). If fluid movement is brought to a complete halt at the head of the tube, then according to Bernoulli, its pressure would rise by $\frac{1}{2}\rho U^2$ and this would be recorded by the manometer to provide a measure of U. The quantity $\frac{1}{2}\rho U^2$ is known as the *dynamic pressure* and is frequently included in expressions describing fluid behaviour.

An intriguing biological example of the Bernoulli effect has been suggested by Vogel and Bretz (1972) to explain the ventilation of the underground burrow of the prairie-dog. They draw attention to the animal's habit of building a mound at one end of its tunnel to raise the opening above ground, leaving the other opening at ground level. Since the wind velocity at the top of the mound is higher than that at the ground surface, there will be a difference in air pressure between the two ends causing air to circulate through the tunnel. The authors calculate that with a wind-speed of $0\cdot2$ m s^{-1} above one opening and $0\cdot1$ m s^{-1} above the other, the air in a typical burrow would be renewed every 10 minutes. A similar explanation is given for the unidirectional flow of water through the burrows of intertidal worms (e.g. the lugworm) which have conical mounds at one end and a depression at the other.

As will be explained later, differences in fluid velocity and hence in pressure are also developed over the curved surface of a body when moving in a fluid. It has been shown experimentally (Miller 1966) that for certain large beetles in simulated flight, sufficient pressure differences can be established between the spiracles that line the sides of the body to provide essential extra ventilation to the flight muscles via the trachea.

Pressure drag

It is common knowledge that a body immersed in a fluid is subject to a fluid pressure acting normal to its surface. Fig. 5.3a illustrates the situation around a very long vertical cylinder lying stationary in a fluid; under these

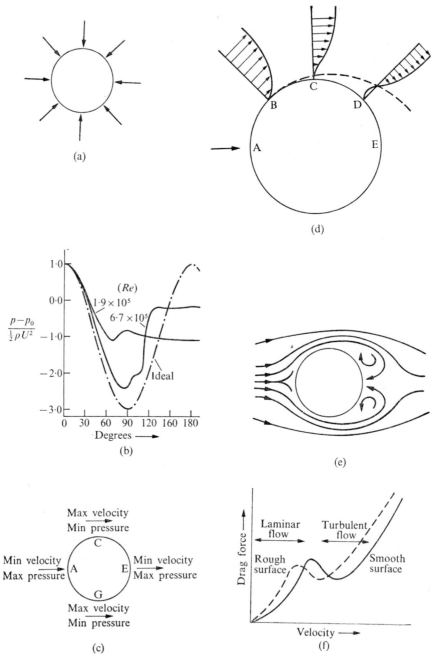

FIG. 5.3. Flow and pressure distribution around a cylinder; (a) static pressure distribution; (b) pressure distribution around a cylinder in an ideal and in a viscous fluid moving at right angles to the axis; (c) velocity–pressure relationship around cylinder; (d) boundary layer separation; (e) turbulent wake behind cylinder; (f) drag on cylinder as a function of fluid velocity and surface roughness.

circumstances the fluid pressure will be uniform around the perimeter, as indicated by the arrows. If the cylinder is now moved along a line at right angles to its axis, or if the fluid moves over the cylinder in the opposite direction, the distribution of the fluid pressure around the cylinder is changed. In the case of an ideal, i.e. non-viscous fluid, it can be deduced theoretically that the pressure distribution will be as indicated in Fig. 5.3b, with points of maximum fluid pressure at the front and rear, and minimum pressures half way round. The explanation is as follows: at A (Fig. 5.3c), the point directly in line with the flow, fluid movement is brought to a halt (the so-called *stagnation point*); as the fluid is diverted to flow over the front half of the cylinder, it accelerates to reach a maximum velocity half way round, at C and G. Over the rear half of the cylinder the fluid then slows down to a second stagnation point at the rear, E and then increases its speed once more behind the cylinder, eventually attaining the initial free stream velocity. It follows from Bernoulli, therefore, that the fluid pressure on the cylinder will be maximum at A and E, and minimum at C and G. The pressure distribution around the cylinder in Fig. 5.3b is presented as the dimensionless quantity, $p - p_0/\frac{1}{2}\rho U^2$ where p = pressure at a given point on the cylinder, p_0 = pressure up-stream where the velocity = U and $\frac{1}{2}\rho U^2$ = the dynamic pressure.

An inspection of the ideal pressure distribution shown in Fig. 5.3b shows that the total fluid pressure over the front half of the cylinder, tending to oppose body movement in the fluid, exactly matches that over the rear half which tends to assist movement. An interesting, though perhaps somewhat academic conclusion, is that in the absence of skin friction, the net drag on the cylinder is zero, i.e. it can move through the fluid without any resistance.

All fluids of course, possess some viscosity so that as they flow over the surface, a boundary layer is formed and because of skin friction, fluid near the surface becomes progressively slowed down. With very small bodies, moving at slow speeds and/or in highly viscous media, i.e. with very small Reynolds' numbers, the boundary layer is so thick that the effects of pressure changes in the surrounding bulk fluid can be neglected; in such cases the drag may be assumed to be entirely due to skin friction.

With larger bodies, moving at higher velocities and thus with higher Reynolds' numbers, the boundary layer is much thinner and flow within it more susceptible to changes in fluid pressure outside. Over the front half of the cylinder where the fluid pressure is decreasing and acts on the body in the same direction as the flow, the boundary layer fluid is unaffected, but over the rear half, we have even slower moving fluid near the surface together with an increasing bulk fluid pressure in the opposite direction. The result is that, at some point on the rear surface, fluid movement near the surface is brought to a halt and then begins to move in the opposite direction. This is known as *reversal of flow*.

The situation at different points around the cylinder surface is illustrated by the velocity profiles in the boundary layer (Fig. 5.3d). At B, the velocity profile corresponds to that in a normal laminar boundary layer. At C we begin to see the first sign of reverse flow in the shape of an inflection in the profile next to the surface, whilst at D where reverse flow is well established, the line of zero velocity is located some finite distance from the surface. This is known as *boundary layer separation*. A common consequence of separation is the formation of relatively large eddies in the wake of the body (Fig. 5.3e); of greater significance is the diversion of the main flow away from the body preventing it from regaining its original velocity and pressure, thus producing a region of low pressure immediately behind the cylinder. The actual pressure distribution over the body surface is now as depicted in Fig. 5.3b (solid lines). We see that depending on the Reynolds' number (*Re*), the fluid pressure over the front half is now much greater than that over the rear half with a resultant net force opposing movement. Hence the term pressure drag.

As the speed and/or size of the body increases, separation with eddying becomes more pronounced and the total drag increases. However, when the Reynolds' number exceeds a value of around 10^5 to 10^6 (the lower value applies when the bulk flow is turbulent), boundary layer flow changes from laminar to turbulent. Because turbulence allows momentum to be transported more rapidly from the freely moving bulk fluid to the slowly moving fluid at the surface, this tends to counteract the back pressure. A turbulent boundary layer is therefore much more robust than a laminar boundary layer; the point of separation moves to the rear of the cylinder and there is much less eddying. The result is that despite the increased friction drag associated with turbulent flow (Fig. 4.2), the pressure drag is so much smaller that the total drag actually decreases (Fig. 5.3f).

Much the same happens in the case of short cylindrical bodies and spheres, but the situation is rather more complicated because of the three-dimensional nature of the flow round these bodies.

An interesting situation arises when the surface, say of a sphere is roughened. As explained earlier (p. 25) this promotes the development of turbulence at a lower speed (lower (*Re*)) than in the case of a smooth surface, with the result that over a certain range of velocities, the total drag is much less (Fig. 5.3f). This might explain the superior performance of 'dimpled' as compared with smooth golf balls, an observation which appears to have been made quite empirically long before the reason was known. At low speeds (low (*Re*)) when the drag is largely due to skin friction the reverse is the case.

Streamlining

Bodies such as cylinders and spheres with extensive boundary layer separation and with a wake of large eddies are known as *bluff bodies*, and as we have seen, are usually associated with relatively large drag forces. It is common knowledge that the drag can be substantially reduced by streamlining.

The essential feature of the streamline shape is its long tail (Fig. 5.4). Fluid behaviour over the rounded front or nose is much the same as over the front half of a sphere; the function of the long tail is to produce such a gradual increase in the pressure of the fluid that the laminar boundary layer remains attached over most if not all the surface. Once past the shoulder the fluid decelerates slowly and merges smoothly into the main stream without forming significant eddies (see Fig. 5.4). The result is a much lower pressure

FIG. 5.4. Flow around a streamline body.

drag than for a bluff body of the same size and speed. Skin friction drag will of course, still occur and it will be appreciated that increasing the surface area by undue lengthening of the tail, will increase this drag.

Some quantitative relations

An important objective of fluid mechanics is to evaluate the drag force exerted on a body either when the body moves relative to a fluid or vice versa. The classic work in this field was done by Rayleigh and was based on the argument that for a body of given shape and attitude the drag force must depend on (1) the relevant physical properties of the fluid, i.e. its density ρ and viscosity, η (2) the speed of the fluid relative to the body (U) and (3) the size of the body. Any linear dimension (l) such as length or diameter can be used as a measure of body size, depending on which is most convenient.

The problem is then one suited to dimensional analysis i.e. to group the above variables in such a way that the combined product represents a force (the drag force) with dimensions MLT^{-2}. This cannot be solved completely because of certain unknowns, but if these are allowed for in the form of a factor (a pure number), it can be shown that:

$$\text{drag force} = (\text{number}) \times \tfrac{1}{2}\rho U^2 l^2 \left(\frac{\eta}{Ul\rho}\right)^c.$$

where c is another number.

We can recognise in this expression the dynamic pressure term $\tfrac{1}{2}\rho U^2$ and the inverse of the Reynolds' number, $Ul\rho/\eta$. The term l^2 then represents an area A and the choice of this area largely depends on convention; for a sphere, cylinder or streamline body, A is taken as the maximum cross-sectional area presented by the body perpendicular to the line of flow. For a sphere of radius

r, A will be the area of its median plane, πr^2. The general expression then reduces to:

$$\text{Drag force, } F_D = \tfrac{1}{2}\rho U^2 A f(Re), \tag{5.2}$$

where f signifies a 'function of'. Since (Re) is a pure number so is $f(Re)$.

Eqn (5.2) is of fundamental importance not only in clarifying the behaviour of a fluid around an immersed body, but also in simplifying the presentation of experimental data. If $f(Re)$ is the same for two bodies (irrespective of whatever combination of U, l, η gives this number) the drag force for a given value of ρ and U, will be solely determined by A. This is known as the principle of *dynamic similarity* and underlies the practice of predicting the drag on large bodies from measurements made on smaller models, as for example, in wind tunnels.

In presenting experimental data, it is customary to express the drag force per unit area, F_D/A in terms of a corresponding drag coefficient C_D defined as before by:

$$C_D = \frac{F_D}{\tfrac{1}{2}\rho U^2 A} = f(Re). \tag{5.3}$$

In other words, for a body of given shape and attitude all that is required is to express C_D as a function of (Re). Data for variously shaped bodies are presented in Fig. 5.5 on a log–log scale.

For very small spherical bodies of diameter d, Stokes showed that:

$$F_D = 3\pi d\eta U. \tag{5.4}$$

This equation, known as Stokes' law, corresponds to that part of the curve for a sphere in Fig. 5.5 when (Re) is less than 1, and as explained in more

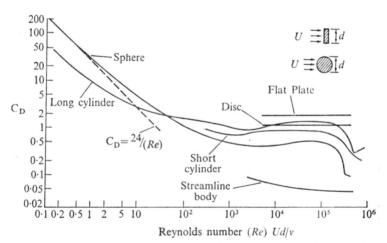

FIG. 5.5. Drag coefficients as a function of the Reynolds' number (Re) for variously shaped bodies. Data of Eisner (1930), Prandtl and Tietjens (1934) and Goldstein (1938).

detail in Appendix 7, underlies the experimental determination of the viscosity of liquids, and/or the equivalent diameter of small particles. From eqn (5.3) with $A = \pi d^2/4$, it can easily be shown that $C_D = 24/(Re)$.

At these very low Reynolds' numbers, a laminar boundary layer persists over the whole body without separation so that the drag is entirely due to skin friction, i.e. it is entirely viscous drag. The fact that the drag is then directly proportional to the diameter and speed of the body has important biological consequences. If the diameter of a small spherical organism is halved, the drag which it has to overcome in swimming is also halved. The energy available, however, will be approximately proportional to its volume and halving the diameter will mean a reduction of about 2^3 or 8 times the amount of available energy. This can only be accommodated by a corresponding reduction in its speed. In other words, the smaller the organisms, the lower will be the speed it can maintain. Larger organisms will always have an advantage because an increase in size will mean a relatively greater increase in their volume and available energy. However, an increase in body size and speed means an increase in (Re) and when this exceeds 1, we see from Fig. 5.5 that eqn (5.4) no longer holds.

With increasing (Re), Fig. 5.5 indicates that C_D for a sphere falls progressively to a more or less constant value of about 0·5 between (Re) 10^3 to 10^5. Within this range, laminar flow occurs over the front half of the body and separates over the rear half with the formation of a relatively large wake. Pressure drag therefore predominates and the drag force is now more or less proportional to the square of the velocity U^2. Above about (Re) 10^5, corresponding to the formation of a turbulent boundary layer, C_D falls sharply to less than half its previous level and then remains more or less constant.

Much the same sequence is shown by an 'infinitely' long cylinder perpendicular to the line of flow. For shorter cylinders aligned vertically to flow, C_D decreases to approximately half that for an 'infinitely' long cylinder as the ratio of length to diameter decreases. This is because the fluid can escape more and more easily around the ends of the cylinder into regions of low pressure behind, so reducing the pressure drag. The curve represented in Fig. 5.5 as a short cylinder has a length five times its diameter.

Also included in Fig. 5.5 are the C_D values for a circular disc and square plate with their faces perpendicular to the line of flow. We have already seen that when a thin flat plate is aligned parallel to the flow, the drag is entirely due to skin friction (eqns 4.2, 4.6). Even when the plate is inclined to the flow path the drag is still dominated by viscous forces if (Re) is low. At high (Re) however, corresponding to higher speeds and/or larger plates, separation of the boundary layer occurs at the edges, so inducing pressure or form drag. The flow pattern for an inclined plate is illustrated in Fig. 5.6. Streams of fluid impinging on the leading surface will flow along it producing a boundary layer and some skin friction; separation occurs, especially at the upper edge,

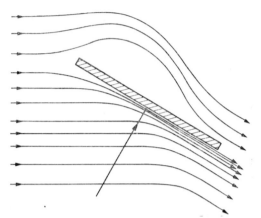

Fig. 5.6. Flow pattern around an inclined plate. Arrow indicates resultant force on plate normal to the surface.

and a turbulent wake is formed behind the plate. This becomes more pronounced as the angle of incidence increases and reaches a maximum when the plate is perpendicular to the flow. Above about $(Re) = 10^3$, C_D for a perpendicular plate becomes independent of (Re) as shown in Fig. 5.5 and therefore, directly proportional to U^2. This has been confirmed by Thom (1968) for an artificial metal leaf held at three different angles (0, 23°, and 90°) in an air stream varying in velocity from 0·2 to 1·5 m s^{-1}; the range of (Re) was $0·5 \times 10^3$ to $3·5 \times 10^3$ and flow can be assumed laminar. At 1 m s^{-1} and above $(Re > 10^3)$, C_D was more or less independent of (Re) with an average value of about 0·44 when the leaf was perpendicular to flow. At $(Re) = 0·5 \times 10^3$, C_D was somewhat higher, with an average value of about 0·5.

The effect of body angle (i.e. attitude) on the pattern of flow is beautifully shown by the photographs of Nachtigall and Bilo (1965) for the water beetle, *Dytiscus* (Fig. 5.7).

An important consequence of fluid behaviour around an inclined surface is that the fluid pressure on the leading surface is increased and that on the trailing surface is decreased; this produces a net positive force on the former tending to move the plate at right angles to the line of flow or movement (Fig. 5.6). The difference in pressure varies with the speed of the plate relative to the fluid and with the angle of incidence; it also varies at different points along the surface so that it is usually difficult to predict its intensity. The significance of this will be discussed in the following chapter on locomotion.

The effect of elongating an ellipsoid shape, leading finally to the streamline form is illustrated in Fig. 5.8 which also contrasts laminar and turbulent flow in the bulk fluid. As the length of the axis in line with flow increases relative to the axis at right angles (this is known as the *fineness ratio*) so C_D decreases, with a minimum value for the streamline shape b (in this case an airship model). We can again recognize the sharp fall in C_D when flow changes from

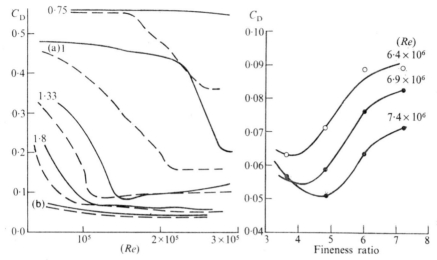

FIG. 5.8. Drag coefficients as a function of Reynolds' number (*Re*) for rotationally symmetrical bodies with different fineness ratios. Full lines—uniform wind, pecked lines—turbulent wind. From data of Prandtl and Tietjens (1934). (a) = sphere, (b) = airship model.

FIG. 5.9. Drag coefficients as a function of fineness ratio for streamline bodies at different Reynolds' numbers. From Goldstein (1938).

laminar to turbulent, especially when the bulk fluid flow is laminar; also the fall in the critical (*Re*) when the bulk flow is turbulent.

In many respects fluid behaviour over a streamline body is similar to that over a flat plate, with laminar flow over the front part and then a transition to turbulent flow when (*Re*) exceeds a value of 10^6 to 10^7 (lower end of the range for turbulent bulk flow). For most practical purposes therefore, C_D for a streamline body may be assumed to be the same as that for a flat plate of the same surface area and the same (*Re*) (based on the axial length of the streamline body), (eqns 4.2, 4.6). This assumes that all the drag on a streamline body is frictional, but depending on the fineness ratio, some form drag may still occur. Fig. 5.9, which illustrates the effect of the fineness ratio of a streamline body on its C_D suggests a minimum at a ratio of about 5. Such a minimum is not unexpected since if the ratio is low, as with a sphere, separation will occur and the drag will be high, whereas if the ratio is high, the increased surface area will mean a higher skin friction. It is significant that the dolphin body has a fineness ratio of about 5, thus very near the optimum.

The story of the dolphin

Dolphins have aroused particular interest since certain species at least have been credited with sustained speeds of 9 ms^{-1} or even more, therefore reflecting a very high efficiency (Kramer 1965; Lang and Pryor 1966).

According to Kramer, the white-bellied dolphin appears to experience only about $\frac{1}{5}$th of the drag of a streamline torpedo of similar displacement (Fig. 5.10). His calculations are as follows.

For the torpedo with body length, $l = 2\cdot16$ m, moving at a speed U of $9\cdot27$ m s^{-1} in sea water ($v = 1\cdot07\times10^{-6}$ m² s^{-1}), $(Re) = Ul/v = 1\cdot87\times10^7$. With a critical (Re) of 5×10^5 flow will be turbulent over most of the body length.

We may assume that a streamline body can be treated as a flat plate with the same surface area.

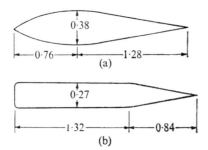

FIG. 5.10. Body shapes and dimensions (m) of (a) white-belly dolphin and (b) a modern torpedo of the same displacement. Data of Kramer (1965).

The density of sea water, $\rho = 1\cdot02\times10^3$ kg m^{-3} and the total wetted surface area $A = 1\cdot56$ m². Hence from eqn (4.7) the total drag force, $F_{DT} = 175$ N.

The power required to propel the torpedo at the given speed of $9\cdot27$ m s^{-1}

$$= F_{DT}\times\text{speed} = 1\cdot62\times10^3 \text{ W (about 2 H.P.)}.$$

Kramer estimated the sustained power output of a dolphin at about $3\cdot3$ watts per kg body weight, about three times that of a trained human being. For the dolphin illustrated in Fig. 5.10, with a body weight of 93 kg, this would mean a power output of about 307 watts. This is about $\frac{1}{5}$th the power required to propel the torpedo at the same speed and even then assumes that all the dolphin's power output is available for propulsion.

Comparing eqns (4.3, 4.7) at $(Re) = 2\times10^7$, the drag with laminar flow is about 10 times less than that with turbulent flow and attempts have been made to explain the efficiency of the dolphin on this basis. Kramer has suggested that the special coating on the skin might stabilize laminar flow by damping down fluid oscillations at the surface.

From observations of trained specimens of *Stenella*, however, Lang and Pryor (1966) calculated that during coasting, the drag coefficient was the same as that of an equivalent rigid body with a near-turbulent boundary layer. They estimated that per unit body weight, the maximum power output was 50 per cent greater than that of a human athlete.

Fish schooling

In a particularly interesting application of the principles of hydro-mechanics, Weihs (1973a) has attempted to explain the advantages of the schooling habit of fish. The vortex (eddy) trail left by a fish swimming in water induces a water movement opposite to the swimming direction immediately behind the fish, but in the direction of swimming to the side. A fish swimming immediately behind another would therefore experience a greater drag than if it swam midway behind two others. The resulting diamond pattern of schooling would then be consistent with a smaller individual effort.

6. Animal Locomotion

ANIMAL locomotion reveals a wide variety of mechanisms involving several different aspects of fluid behaviour. Most of the underlying principles have already been discussed and the object of this chapter is to apply these principles to specific examples. Though some reference will be made to energy relations and efficiencies, little will be said about mechanical and physiological relations; for these and other aspects, the reader is referred to several specialist publications including Chadwick (1953), Pringle (1957), Brown (1963), Gray (1968), Alexander (1968), Hertl (1968), and Pennycuick (1972).

Rowing

Probably the simplest mechanism of locomotion is rowing, as found in the majority of the ciliate protozoa. The diameter of a cilium is about 0·2 μm and its length varies from about 5 to 10 μm, so that the rowing action takes place at very low Reynolds' numbers when only viscous forces will be involved. The small size of the organisms and their relatively low speeds means that the drag on their bodies will also be viscous and hence directly proportional to their speeds (p. 53).

For the power stroke (Fig. 6.1a), each cilium is rigid and as it beats through the water, the opposing drag causes the organism as a whole to move in the opposite direction. The underlying principle is described in Newton's third law of motion, namely that 'action and reaction are equal and opposite'. On the return stroke (Fig. 6.1b) the cilia are flexed to reduce the reverse thrust.

In most ciliates, the cilia are arranged in longitudinal rows. In some the beat is simultaneous, in others the action is so synchronized that whilst some are in the power stroke, others are on the return stroke; the wave of ciliary movement so produced (rather like a field of corn in the wind) is called the *metachronal wave*. Though of much physiological interest because of the mechanism of synchronization, this wave probably has no mechanical significance other than to produce a smooth forward movement.

In multicellular organisms beating of the cilia is used for a variety of specialized purposes, e.g. in ciliated epithelia the cilia propel nutrients, egg cells, foreign bodies etc. across the surface.

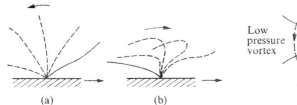

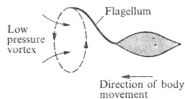

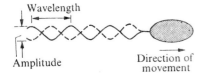

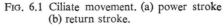

FIG. 6.1

FIG. 6.2

FIG. 6.3

FIG. 6.1 Ciliate movement. (a) power stroke (b) return stroke.

FIG. 6.2. Flagellum action in *Euglena*.

FIG. 6.3. Propulsion by undulatory movement of a flagellum.

The rowing mechanism is also found in larger organisms. Water beetles propel themselves under water by a rowing action of the middle and hind legs which are so jointed as to enable the tarsus to be rotated and feathered like an oar in the return stroke. Working as they do at much higher Reynolds' numbers than protozoan cilia, only inertial forces need be considered. Depending on the size of the legs, their speed and attitude, a certain mass of water is accelerated backwards and the reactive force of the water on the legs drives the whole body forward, exactly as in rowing a boat.

Nachtigall and his colleagues have studied in great detail the swimming mechanism in *Dytiscus* and *Acilius* and the results, together with an account of the action of other swimming insects, are reviewed by Nachtigall (1964). The bodies of most water beetles are dorso–ventrally flattened and when oriented parallel to their line of movement, they correspond reasonably well to streamline bodies with drag coefficients ranging from about 0·23 to 0·35 (see Fig. 5.5). In the case of *Dytiscus*, with a body length of 0·03 m, swimming in water ($v = 10^{-6}$ m² s⁻¹) at speeds ranging from 0·1 to 0·4 m s⁻¹, the Reynolds' number will vary from about 3×10^3 to $1·2 \times 10^4$; according to Fig. 5.5, the drag is only about 3 to 5 times that on a perfectly streamlined body.

Increasing the vertical inclination of the body increases the drag up to about 9 times at 90°, reflecting the increasing boundary layer separation, as is clearly illustrated in Fig. 5.7.

As shown in Appendix 8, the efficiency of the rowing action, i.e. the ratio of the mechanical power achieved by body movement in the fluid to that of the total power expended in rowing, is determined by the ratio of the forward speed of the body U to the speed of the 'oars' u, and by the drag forces, f and

f' on the 'oars' during the power and return strokes respectively:

$$\text{efficiency} = \frac{U(f-f')}{u(f+f')} \, . \tag{6.1}$$

It follows from eqn (6.1) that the efficiency will be increased by lowering the speed of the oars relative to that of the body and by decreasing the drag on the return stroke. The forward speed of the body will of course be determined by the stroke frequency which in water beetles, ranges from about 2 to 5 per second in the larger species (e.g. *Dytiscus*) to about 16 to 20 per second in the smallest. In general however, the speed of the leg rowing action is of the same order as that of the body speed. The legs of *Dytiscus* are flattened in cross-section so that with the broader surface presented to the water during the power stroke and the narrower surface during the return stroke, the drag coefficient for the former (about 1·29) is very much larger than that for the latter (about 0·03). In *Acilius*, the legs are thickly covered with hairs which spread out during the power stroke and lie flat during the return stroke; the result is again, a higher drag coefficient in the power stroke than in the return stroke. The same effect is achieved in rowing a boat by using large blades and feathering on the return stroke.

If f is large compared with f', the latter can be ignored so that the efficiency will then be determined mainly by the ratio U/u.

Flagellum action

Several methods of propulsion occur in the Flagellata, some of them rather unique (Jahn and Votta 1972). The flagellum appears to be similar in structure to a cilium, but is thicker and has a length of up to about 150 μm. Again, at the very low Reynolds' numbers prevailing (for spermatozoa, (Re) is about 10^{-3}), the organisms experience only viscous forces.

Euglena achieves most of its propulsion by rotating its single flagellum. As indicated in Fig. 6.2, this creates a vortex of low fluid pressure at the anterior end (rather like the vortex produced when a cup of tea is rapidly stirred) and it is possible that the organism is propelled by the higher fluid pressure behind. As a consequence of flagellum movement the body rotates in the opposite direction.

Among most of the flagellates however, propulsion is obtained by a characteristic bending and straightening of the flagellum which appears as an undulatory wave usually travelling from the point of attachment to the tip (Fig. 6.3). This mechanism has been analysed by Taylor (1951, 1952a) and Hancock (1953) who showed that when the amplitude of the undulations was small compared with their wavelength, the speed of the organism was proportional to the speed of the waves along the flagellum and to the square of their amplitude, and inversely proportional to the square of the wavelength. If the waves are propagated in spirals, but with the same amplitude, the speed of forward movement will be twice that of wave motion in a single plane.

Fish propulsion

Depending on the relative length and flexibility of the tail, three main types of propulsion have been recognized (Bainbridge, 1963). These are:

    (1) *anguilliform* types, with long flexible tails like the eel (Fig. 6.4a);

    (2) *carangiform* types, with tapering tails of medium length terminating in a well defined caudal fin, as in salmon, whiting and most pelagic species (Fig. 6.4b);

    (3) *ostraciiform* types, with relatively rigid bodies and short tails (Fig. 6.4c).

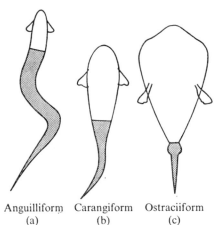

FIG. 6.4. The three main types of caudal propulsion in fish. The portion of the body taking part in oscillatory movement is shaded. From Gray (1968). *Animal Locomotion*. Published by Weidenfeld (Publishers) Ltd, London.

Anguilliform   Carangiform   Ostraciiform
(a)   (b)   (c)

Anguilliform types propel themselves through the water by an undulatory movement produced by alternate contractions of the muscle segments on either side of the body. The resulting waves pass down the body with increasing amplitude towards the tail. Though superficially similar to the undulations of flagellae, fluid behaviour is very different because fish operate at much higher Reynolds' numbers and therefore, inertial forces predominate.

Taylor (1952b) has analysed propulsion by undulatory movement using a simplified model of a flexible cylinder with waves of constant amplitude travelling down the cylinder at constant speed. The forces involved then correspond to those generated by a cylinder moving through water at a constant speed with its axis inclined at an angle to the direction of movement (Fig. 6.5). Using an argument similar to that for an inclined plate (p. 54), the reactive force of the water R normal to the cylinder can be resolved into a propulsive component F driving the body forward, and a component L at right angles to F. Taylor showed that for a given (Re), the forward speed of the

Direction of wave motion ⟶

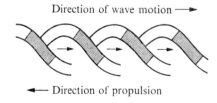

⟵ Direction of propulsion

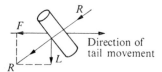

FIG. 6.5. Anguilliform propulsion. Forces generated by a cylinder moving at right angles to the direction of forward movement. R = reactive force of water, F = propulsive component, L = lateral component.

body was determined by the amplitude and wavelength of the undulations. It is of interest that if the surface of the body is very rough, forward propulsion can be achieved by a *forward* movement of the waves, as appears to be the case with certain marine worms. Working on the same problem, Lighthill (1960) showed theoretically, that the maximum efficiency of undulatory propulsion was obtained when the waves travelled at a speed about $\frac{5}{4}$ of that of the swimming speed and when, as in the eel, the amplitude increased from zero at the front to a maximum at the tail.

In the carangiform types, alternate contractions of the muscles on either side of the body again tend to produce an undulatory movement, but because the body is neither so elongated nor as flexible as that of say, an eel, most of the oscillations appear in the caudal fin.

The principle of propulsion is basically the same as that underlying sculling with a single oar from the rear of a boat, but with important modifications arising from changes in the shape of the tail during each stroke. The body movements of the whiting (Gadus) are illustrated in Fig. 6.6a for a single traverse, i.e. a half-beat of the tail. It will be seen that during most of the beat, the posterior region of the body and the caudal fin are directed obliquely backwards. The forces generated as the fin moves through the water are in principle, exactly the same as those generated by an inclined plate (p. 54). As illustrated in Fig. 6.6b the reactive force R normal to the surface can be resolved into a propulsive force F driving the fish forward, and a lateral force L tending to turn the body in a clockwise direction. On the return stroke, the forward thrust is combined with a lateral force tending to turn the body in the opposite direction. Though this alternating lateral movement (so-called *yaw*) is opposed to some extent by the resistance of the water, the path of the fish is typically sinuous (Fig. 6.6c).

The magnitude and direction of the forces acting on the fin are complicated by the forward motion of the fish. The effect is illustrated in Fig. 6.6d; if the sideways velocity of the fin is U_f and the forward velocity of the body is U_b, a triangle of velocities can be constructed to define both the magnitude and direction of the resultant velocity U_R. In other words, the fin 'sees' the water coming from a direction with a velocity depending on the magnitudes of the sideways and forward velocities. This will influence the magnitude of the forces acting on the fin.

Further complications are introduced by dorso–lateral curvature of the fin during its traverse, leading to changes in area, and by changes in the in-clination of the fin to the vertical. The latter, as seen from the rear of the fish, are illustrated in Fig. 6.6a. As the fin approaches the median line, it is rotated about its long axis so that the upper edge leads the lower; after passing

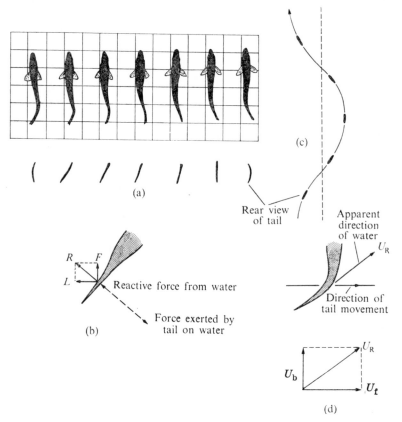

FIG. 6.6. Carangiform propulsion. (a) Body movements of whiting (*Gadus*) during a single traverse (from Gray 1968 *Animal Locomotion*, Weidenfeld (Publishers) Ltd, London). (b) Forces generated by tail movement: R = reactive force of water, F = forward pro-pulsive component, L = lateral component. (c) sinuous pathway of body movement with rear views of tail. (d) triangle of velocities illustrating apparent direction and speed of oncoming water (U_R) being the resultant of tail movement (U_f) and body movement (U_b).

the median line, a counter-rotation brings the fin once more into the vertical so that at the end of the traverse, the attitude of the fin is the same as that at the beginning. These changes in vertical orientation are also illustrated in Fig. 6.6c for different positions during forward progress of the fish. At certain stages therefore, (e.g. position 2) vertical forces will be generated which act as lift forces; in combination with the orientation of the pectoral fins, these forces maintain the horizontal position of the fish in the water. As Bainbridge (1963) has shown, the overall effect of these complex movements is such that at all stages during the traverse, at least some part of the fin presents a positive angle to the water so as to provide a forward thrust.

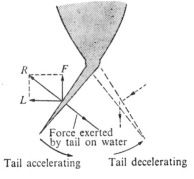

FIG. 6.7. Ostraciiform propulsion. R = reactive force of water on tail, F = forward propulsive component, L = lateral component.

A marked feature of certain carangiform types is their ability to accelerate very rapidly to a high velocity from rest; according to Weihs (1973b), this acceleration is of the order of 40 m s^{-2} for trout and 50 m s^{-2} for pike. He shows that this is achieved by first curling the body into a L shape, followed by a very rapid lunging of the caudal fin and a straightening out of the body; as a result, the fish moves initially at an angle to its original orientation.

In the ostraciiform types (Fig. 6.4c), the short tail oscillates more or less symmetrically about the point of attachment and shows little sign of bending. The method of propulsion reveals distinct differences from that of the carangiform types. During the first half of the traverse (Fig. 6.7) the inclination of the fin is such as to produce a propulsive component F and a lateral component L; the latter is responsible for the sinuous pathway through the water. During the second half of the traverse it might be thought that exactly the reverse forces would be generated, including a negative thrust (indicated by the dotted line). That this does not happen is thought to be due to the fact that during the first half of the traverse, the fin is accelerating, so accelerating the water in front of it. If the water still continues with a high speed during the second half of the traverse when the fin is decelerating, it will still produce a force on the trailing surface of the fin; this will generate a propulsive force in the direction of movement of the fish. It has been argued further that each time the fin reverses its direction at the end of a traverse, it operates against water that has already been set in motion in the opposite direction by the previous sweep. In other words, greater forces are generated than would be the case if the water was initially at rest. It is likely that the same extra propulsive forces are generated in carangiform action.

Energy relations of fish propulsion

Since most fish bodies approximate to the streamline form, the drag imposed on their movement will be largely that due to skin friction, and hence proportional to their surface area, i.e. to a length squared (L^2). As argued earlier (p. 53) the energy available will be roughly proportional to their volume (L^3) so that the longer the body, the greater will be the relative energy available and the higher the speed. For this reason, comparisons between individuals of a species are traditionally based on speeds expressed in terms of body length, the so-called *specific speed*.

For small fish, up to about 0·3 m long, the cruising speed is usually of the order of 4 body lengths per second, though much higher speeds can be reached in short bursts.

It has been established experimentally that within a particular species (dace, trout, goldfish, etc.) the maximum specific speed increases almost linearly with the frequency of the tail beats. For example, with each beat of tail, dace and trout travel a more or less constant distance corresponding to about 60 per cent of their body length. There is however, a limit to which the speed can increase with length because above a certain length, the maximum tail beat frequency begins to decrease. With dace for example, the tail beat frequency increases to a maximum of about 25 strokes per second up to a body length of about 0·07 m and then decreases rather rapidly with further increase in body length (Hertl 1966). It has also been found that the specific speed also depends on the amplitude of the tail beat and that this decreases with increasing tail beat frequency up to about 5 beats per second. Thus, even within a particular species, there are several interacting factors determining their speed.

At a uniform speed (U) the propulsive force must balance the drag force (F_D); the mechanical power output will then be UF_D. F_D may be conveniently determined from the rate of loss of speed when a fish glides in water after a bout of active swimming. Assuming that F_D is proportional to the square of the speed, i.e. $F_D = kU^2$ it can then be shown (Appendix 9) that if the initial velocity is U_0, the distance s travelled in the glide after time t is given by:

$$s = \frac{m}{k} \ln\left(\frac{kU_0 t}{m} + 1\right) \tag{6.2}$$

where m is the mass of the fish.

Cinematographs of small trout (mass $= 0.08$ kg) gave values of $U_0 = 1.4$ m s^{-1} and $k = 1.2$ (Gray 1968); hence, at the beginning of the glide, $F_D = kU_0^2 = 2.4$ (N). Assuming the same drag during active swimming at the same speed, this would mean a mechanical power output of $2.4 \times 1.4 = 3.4$ watts approximately, or about 40 watts per kg body mass. This agrees with other estimates of the power output of fish (usually quoted in terms of body weight instead of mass). With a cross-sectional area of 10 cm^2 (10^{-3} m^2), the drag coefficient for the trout can be calculated from eqn (5.3); substituting $F_D = 2.4$ N, $A = 10^{-3}$ m^2, $\rho = 10^3$ kg m^{-3} and $U = 1.4$ m s^{-1} gives $C_D = 2.5$. Tests on model or dead fish towed at similar speeds have given similar drag coefficients. If the fish was 0.1 m long, its Reynolds' number at the speed of 1.4 m s^{-1} would be about 2×10^3; inspection of Fig. 5.5 suggests that the drag coefficient of 2.5 is somewhat high compared with a truly streamline body.

Flight

Active flight, as distinct from passive gliding, depends on wing action to provide the necessary lift and propulsive forces.

In insects, the wings arise as extensions of the cuticle; they are generally arched or cambered and stiffened by thickened ridges, or veins (Fig. 6.8a). Muscular connections to the sides of the thorax allow the wings to be moved up or down, or rotated.

Bird wings are much more complicated structures. The main lifting surface is provided by the primary feathers, about 10 in number, attached to the hand bones (Fig. 6.8b). The wings are operated by muscles connecting them to the body and can be pivoted upwards or downwards or backwards or forwards; they can also be tilted by rotation of the humerus. In contrast to insect wings, the shape and area of birds wings can be altered by the action of muscles operating at the elbow.

Although in the following accounts, frequent reference will be made to aerodynamic principles underlying the operation of conventional aircraft,

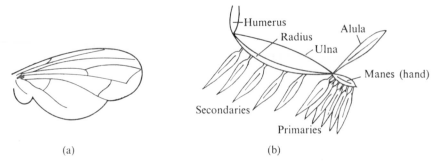

(a)   (b)

FIG. 6.8. (a) Wing of the common blow-fly (*Phormia*). (b) Structure of typical bird's wing.

it must be emphasized that flapping flight is far more complicated and much less well understood than the flight of an aeroplane powered independently of the wings. This becomes quite clear when reading detailed accounts of insect flight (e.g. Nachtigall 1966) and even more so in the case of bird flight (e.g. Brown 1963). It is often difficult to identify precisely, the magnitude and direction of the forces generated during flapping flight and the simplifications introduced here are intended solely to clarify the underlying principles.

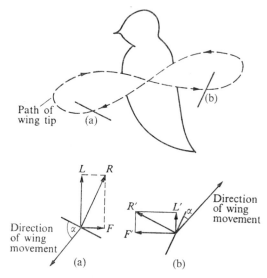

FIG. 6.9. Wing action of hummingbird during hovering with forces diagram (a) during downstroke and (b) during upstroke.

It is perhaps best to begin with hovering flight since this is not complicated by forward propulsion and the action is very similar in insects and birds.

As illustrated in Fig. 6.9, the body is held at a steep angle to the horizontal and the wings are swept rapidly backwards and forwards close to the horizontal plane. At the end of each stroke the wings are rotated so that the wing tip describes a figure-of-eight. The lines in Fig. 6.9 indicate approximately the plane of the wings at point (a) during the down stroke and at point (b) during the upstroke. Because the wing has been rotated in between, the leading surface at (a) will be opposite to that at (b).

At (a), when the wing is moving forward and downward, the component force R, assumed to be generated normal to the wing surface, can be resolved into a vertical force L acting as a lift force and a horizontal force F tending to move the body backwards. At (b), when the wing is moving backward and upward, the normal force can again be resolved into a lifting force L' and a horizontal force F' counteracting the backward movement.

6

The magnitude of the lift force, which must of course balance the weight of the animal, depends on several factors, including the velocity of wing movement, the size of the wings and the angle between the plane of the wing and its direction of movement, the so-called *angle of attack* indicated in the figure by α. These factors will vary throughout the stroke, but basically the action is similar to that of a helicopter rotor blade, except that the wings sweep through an angle of about 120° instead of 360°. An estimate of the power requirements for hovering can in fact be made from a formula derived for the helicopter rotor blade (Prandtl 1952; Alexander 1968).

If P is the minimum power required to maintain a body of mass m stationary in air of density ρ, then:

$$P = \sqrt{\left(\frac{m^3 g^3}{2 A \rho}\right)},\qquad (6.3)$$

where A is the area swept out by the wings and g is the acceleration due to gravity. For air, $\rho = 1 \cdot 3$ kg m^{-3}.

For a small hummingbird of mass 0·003 kg with wings 0·004 m long sweeping over an angle of 120°, the power P works out as $5 \cdot 5 \times 10^{-2}$ W. Measurements of the oxygen uptake of the hummingbird during hovering suggest a total chemical energy output of 0·64 W, assuming complete oxidation of carbohydrate or fat. Only part of this energy of course, is available for mechanical work since much is lost as heat. Assuming that the bird is at least as efficient as a man pedalling a bicycle, for whom the mechanical conversion efficiency is about 20 per cent, this means that about 0·15 W is available for hovering flight, thus giving an aerodynamic efficiency of about 40 per cent (i.e. $5 \cdot 5 \times 10^{-2} \times 100/0 \cdot 15$).

Weis-Fogh (1972) has studied in some detail the energetics of hovering flight in the hummingbird and in the insect *Drosophila* and quotes aerodynamic efficiencies of 50 and 30 per cent respectively. In a later paper (Weis-Fogh 1973), he describes ways of simplifying the estimates of hovering energetics in insects from data on stroke frequency and wing parameters. These include estimates of wing inertia. Whilst the majority of hovering insects appear to depend on normal wing action, as described above, it seems that certain small insects with low Reynolds' numbers (e.g. many butterflies) operate on different, so-called non-steady aerodynamic principles. These involve a rapid twisting of the wings at the beginning and end of the downstroke which it is claimed, sets up vortices that generate enough lift to maintain flight.

Turning to forward flight, the situation, as mentioned earlier, is complicated by the combination of wing flapping with a forward movement of the animal as a whole.

The action is illustrated in Fig. 6.10 for the forward flight of the common blowfly, *Phormia* (Nachtigall 1966). Fig. 6.10a shows high-speed photographs of the wing positions during the downstroke and upstroke and Fig.

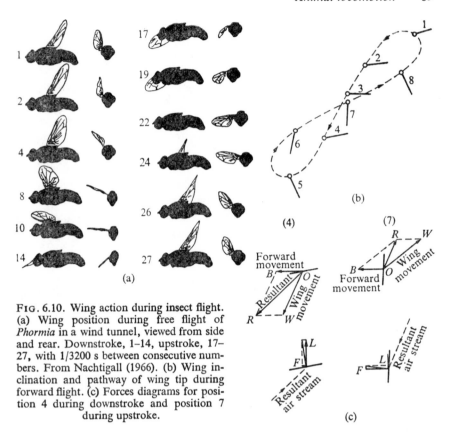

FIG. 6.10. Wing action during insect flight. (a) Wing position during free flight of *Phormia* in a wind tunnel, viewed from side and rear. Downstroke, 1–14, upstroke, 17–27, with 1/3200 s between consecutive numbers. From Nachtigall (1966). (b) Wing inclination and pathway of wing tip during forward flight. (c) Forces diagrams for position 4 during downstroke and position 7 during upstroke.

6.10b the figure-of-eight described by the wing tip, along with the inclination of the wings at different stages. Wing movement is similar to that during hovering, but in a different plane. As before, the magnitude and direction of the resultant air stream which the wing 'sees' can be obtained by constructing the appropriate triangle of velocities. Thus for position 4 during the downstroke (see Fig. 6.10c), the combined effect of a wing velocity in the direction OW and a forward velocity of the insect in the direction OB, gives a resultant velocity OR. The reactive force of the air acts in the opposite direction RO against the underside of the wing, producing a force R normal to the wing surface. As shown in the second diagram, this force may be resolved into a small propulsive component F, and a relatively large lift component L directed vertically upwards. During the upstroke, at position 7, the resultant air stream OR is directed in such a way that its component force R normal to the surface can still be resolved into a substantial propulsive component F and a small lift component L in the right directions.

Detailed studies of the wing orientation show how the angle of attack varies constantly with the flight path (see also Alexander 1968).

Depending on species and speed, wing movement during the fast forward flight of birds shows considerable variation. In the case of the pigeon flying at about 9 m s^{-1} (20 miles per hour), we see from Fig. 6.11(I) that the downstroke begins with the wings fully extended and raised above the body. They are then accelerated downwards and slightly forwards to produce a powerful propulsive force and a small lift force. During the upstroke the wings are flexed to reduce the drag and so generate hardly any propulsive force, though possibly a small lift force. The bastard wing or alula is opened during the downstroke to provide extra lift. Raspet (1961) describes a simple experiment which demonstrates the function of the alula. If a hand is put out of the window of a car travelling at about 50 miles per hour with the palm slightly cupped, and at a positive angle to the wind, opening and closing the thumb will cause a large change in the lifting force on the arm.

A distinctive feature of bird flight is the speed at which many birds can take off from the ground to achieve a relatively high horizontal speed. Much more muscular effort is required of course, than that for forward flight and the pattern of wing movement is very different, as illustrated in Fig. 6.11(II) for the pigeon.

The action begins with the wings fully extended above the body with their dorsal surfaces facing each other (A). Then with a powerful stroke, the wings are accelerated downwards with the alula opened (B–C). After passing the horizontal, the wings are swung forwards from the shoulder joint (CD) so that at the end of the stroke, their under surfaces face each other well in front of the body (D). The main effect of this first stroke is to generate a high lift force without much forward propulsion, as indicated by the inset diagram. The return stroke begins with the wings flexed and drawn above the body so that the primary feathers are directed forwards (E). The wings are then extended and at the same time drawn violently backwards and upwards with the under surfaces leading (F–G). As indicated by the forces diagram corresponding to this stage, a further lift force is generated, but now also some forward propulsion. The upstroke ends with the wings in the same elevated position as at the beginning of the downstroke (H).

Lift–drag ratios

For all animals during steady forward flight, the aim of wing action is to provide a lift force sufficient to balance the weight of the body and a propulsive force sufficient to balance the drag. Most of the underlying principles have been derived from aerodynamic research on fixed aeroplane wings moving at high speeds and considerable attention has been given to wing efficiency defined in terms of the lift/drag ratio. This ratio is usually expressed as the ratio of the lift coefficient, C_L to the drag coefficient, C_D.

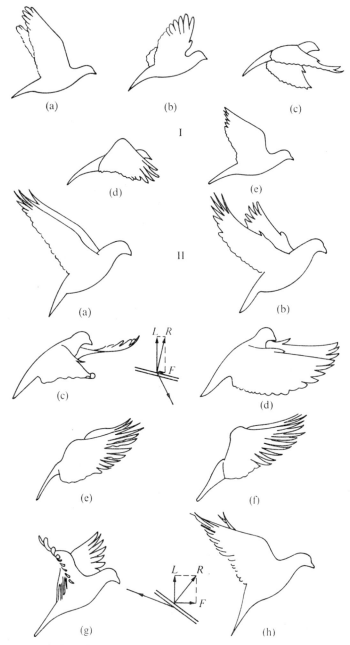

FIG. 6.11. Bird flight. Wing movement of pigeon, (I) during fast forward flight; down-stroke, (a)–(c), upstroke, (d)–(e), (II) during slow flight; downstroke, (a)–(d), upstroke, (e)–(h). Time interval between positions, 1/100 s. Forces diagrams inset for positions, IIC and IIG. *L* = lift component, *F* = forward propulsive component of reactive force *R* on wing. From Brown (1963).

As explained earlier, the drag coefficient is defined as:

$$C_D = \frac{F_D}{\frac{1}{2}\rho U^2 A},\qquad(6.4)$$

where F_D is the drag force, $\frac{1}{2}\rho U^2$ the familar dynamic pressure, and A corresponds to the area of the wings.

By analogy, the lift coefficient C_L is defined as:

$$C_L = \frac{F_L}{\frac{1}{2}\rho U^2 A},\qquad(6.5)$$

where F_L is the lift force and the other terms the same as for C_D.

Aeroplane wings are typically aerofoil in shape and under the right conditions, they give a much higher lift/drag ratio than flat or cambered plates. The aerofoil (Fig. 6.12a) is characterized by a smoothly rounded nose and a

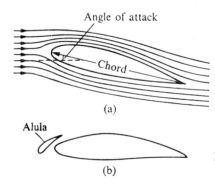

Angle of attack

Chord

Alula

(a)

(b)

FIG. 6.12. (a) Typical aerofoil showing stream lining (b) profile of pigeon wing.

smooth gently curving upper surface, finishing up with a sharp tail. The smooth flow of air over the aerofoil without boundary layer separation (at least at small angles of attack) means a relatively low drag. The profile of a pigeon wing is added for comparison (Fig. 6.12b) to show that in certain birds at least, the wing appears to have evolved towards a similar aerofoil shape. In aeronautical practice, the wing may be slotted near the leading edge so as to lead air from below the wing to the top surface thus reinforcing the boundary layer and reducing the danger of separation at high angles of attack. It is of interest that the slot between the alula and the primary feathers may serve the same purpose, though the same effect can be obtained by separation of the primary feathers.

Despite the enormous practical importance of the aerofoil profile it must be emphasized that its advantages are only to be obtained at high Reynolds' numbers. An aeroplane wing at normal air speeds will have an (Re) (based on the chord) of about 10^7 or higher, whilst a pigeon wing with a chord of about 0·12 m, moving at a speed of 15 m s^{-1}, will have an (Re) of $1·2 \times 10^5$.

By contrast, a locust wing (chord, 0·02 m, speed 3 m s⁻¹) has an (Re) of about 4×10^3 and the wing of *Drosophila* (a few mm long and speed 1 m s⁻¹) an (Re) of about 200. Experiments have shown that the conventional aerofoil is less efficient than a flat or cambered plate at low Reynolds' numbers, so that except in the case of large birds, the aerofoil profile may be a disadvantage rather than an advantage.

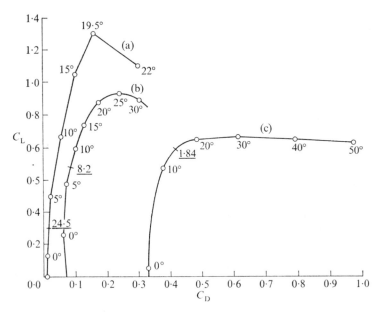

FIG. 6.13. Polar diagrams for (a) conventional aerofoil with (Re) ≈ 5×10^6, (b) locust hind wing with (Re) ≈ 4×10^3, (c) *Drosophilia* wing with (Re) ≈ 2×10^2 at different angles of attack. The point of maximum lift/drag ratio is located on each curve and the value of the ratio at that point is underlined. From Vogel (1967).

Some particularly interesting comparisons have been made by Vogel (1967) of the performance of a *Drosophila* wing (at (Re) 200), the hind wing of a locust (at (Re) 4000) and a conventional aerofoil (at (Re) 5×10^6) for various angles of attack. The data are presented in Fig. 6.13 in the form of a so-called *polar diagram* in which the lift coefficient C_L is plotted against the drag coefficient C_D.

For the conventional aerofoil we see that up to a certain angle, C_L increases with the angle of attack, but as C_D increases more rapidly, the lift/drag ratio begins to fall off. Above a certain angle, C_L falls rapidly, as also the C_L/C_D ratio. This is known as the *stalling angle* and corresponds to separation of the boundary layer from the surface (see p. 50). The highest C_L/C_D ratio of about 24·5 occurs at an angle of about 2° (the point at which the tangent passes through the origin). Though the efficiency of the locust wing is much lower (highest C_L/C_D ratio of 8·2 at 7°), stalling occurs at a rather higher angle of attack than with the aerofoil. The efficiency of the *Drosophila* wing is even smaller, (maximum $C_L/C_D = 1·8$) but in contrast to the others, it appears almost impossible to stall; it has been suggested that this may be related to the presence of hairs or bristles on the wing.

It must be borne in mind that the properties of a wing intended to generate both lift and propulsive forces may be different from those of a fixed wing designed to produce a maximum lift/drag ratio.

Drag during flight

The total drag on an animal flying through the air is made up of the drag on both body and wings. If the body approximates to a streamline form, then except at very low Reynolds' numbers, the drag will be approximately proportional to the square of the flying speed. The drag on the wings, known as *profile drag*, will depend on the shape and area of the wings and also on their attitude, i.e. the angle of attack. Profile drag is also approximately proportional to the square of the speed. However, when a wing moves in air and generates lift, an extra drag arises called *induced drag*. This is attributed to the formation of eddies (vortices) at the wing tips. Induced drag decreases with the wing-span according to the approximate formula:

$$Induced\ drag = \frac{1}{\pi} \cdot \frac{1}{\frac{1}{2}\rho U^2} \left(\frac{\text{Lift}}{\text{Span}}\right)^2. \tag{6.6}$$

In other words, for a given lift/span ratio (since the lift must balance the weight, this ratio is known as *span loading*), the induced drag is *inversely* proportional to the square of the speed. This means that at low speeds the induced drag will be relatively high; in order to reduce this, hovering and slow-flying birds have exceptionally large wing-spans.

In winged flight, the total drag may be expressed as:

Total drag = body plus *profile drag* ($\propto U^2$) plus *induced drag* ($\propto 1/U^2$)

If these are plotted against flying speed as in Fig. 6.14 we see that at a particular speed (U_{md}) the drag is a minimum. It can be shown (Appendix 10) that this occurs when the body plus profile drag equals the induced drag, and the magnitude of U_{md} is then given by:

$$U_{md} = \sqrt{\left(\frac{W}{\frac{1}{2}\rho A C_L}\right)} = \left(\frac{W}{\frac{1}{2}\rho A}\right)^{\frac{1}{2}} \left(\frac{k}{C_{D0}}\right)^{\frac{1}{4}} \tag{6.7}$$

where W corresponds to the weight of the animal, ρ to the air density, A to the wing area, C_L to the lift coefficient, C_{D0} to the drag coefficient for body plus wings at zero lift and k to the ratio $A/\pi b^2$ where b is the wing span.

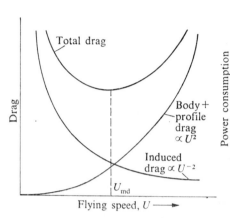

FIG. 6.14. Variation in *body* plus *profile drag*, *induced drag*, and *total drag* with flying speed. U_{md} = flying speed for minimum total drag.

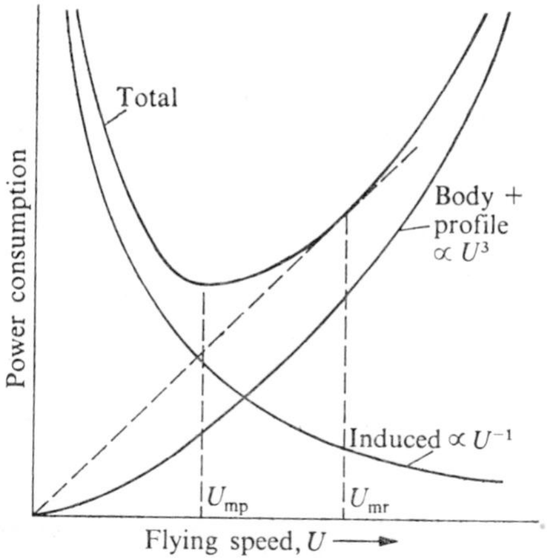

FIG. 6.15. Variation in power required to overcome *body* plus *profile drag*, *induced drag* and *total drag* with flying speed. U_{mp} = flying speed for minimum power consumption, U_{mr} for maximum range.

In other words, the speed corresponding to minimum drag is proportional to the square root of the weight and inversely proportional to the square root of the lift coefficient at that speed.

The power required to maintain an animal in steady level flight is equal to the product of the drag force and the speed. In the case of body plus profile drag, the power will therefore be proportional to U^3, whereas for induced drag, it will be proportional to $1/U$. These two quantities are plotted in Fig. 6.15 and again we see that at a certain speed U_{mp}, the total power required is a minimum. At this speed therefore, a bird can remain airborne for the longest period on a given amount of food energy. However, if as during migration, the aim is to cover the longest distance, the most efficient utilization of food energy is obtained at a somewhat higher speed U_{mr}, corresponding to the minimum power/speed ratio; this is readily located by the tangent from the origin. It can be shown (Appendix 11) that $U_{mp} = 0.76U_{md}$, the speed corresponding to minimum drag as given in eqn (6.7). It turns out that the speed for maximum range (U_{mr}) can never be less than 1·32 times the minimum power speed (U_{mp}) and according to Pennycuick (1972), is more probably 1·8 times. Further details of these relationships are given in Appendix 11.

The effect of size on speed and power requirements can be deduced starting with the definition of the lift force (F_L) in terms of its lift coefficient (C_L) as given in eqn (6.5):

$$F_L = C_L \tfrac{1}{2}\rho U^2 A.$$

This gives:

$$U^2 = \frac{F_L}{\frac{1}{2}\rho A C_L} \quad \text{or} \quad U = \sqrt{\left(\frac{F_L}{\frac{1}{2}\rho A C_L}\right)}. \tag{6.8}$$

Since in level flight the lift force balances the weight W, we have:

$$U = \sqrt{\left(\frac{2W}{\rho A C_L}\right)}. \tag{6.9}$$

Assuming a constant lift coefficient and air density (ρ) it follows from eqn (6.9) that the speed of an animal will be proportional to the square root of the wing loading factor W/A.

The weight of an animal is proportional to the cube of a linear dimension (L^3), whilst the wing area is proportional to L^2; hence W/A will be determined by a length (L). It follows from eqn (6.9) that the speed will be proportional to $L^{\frac{1}{2}}$. If we take L as representing the wing-span, then for animals of the same shape, one with a wing-span say N times that of another should travel $\sqrt{(N)}$ times as fast.

Since L is proportional to $W^{\frac{1}{3}}$, we can express the speed in terms of the body weight:

$$\text{Speed} \propto (W^{\frac{1}{3}})^{\frac{1}{2}} \propto (W)^{\frac{1}{6}}. \tag{6.10}$$

For a given lift/drag ratio, the drag force will be proportional to the lift and hence to the weight W which in turn is proportional to L^3. The power required at a speed U is given by the product of the drag force and U, and since U is proportional to $L^{\frac{1}{2}}$, the power will be proportional to $L^3 L^{\frac{1}{2}}$ i.e. $L^{3.5}$. Alternatively, since L is proportional to $W^{\frac{1}{3}}$, the power required will be proportional to $(W^{\frac{1}{3}})^{3.5}$ i.e. to $(W)^{1.17}$.

It follows from the above that a bird A weighing say, twice as much as a bird B, will require $2^{1.17}$ or 2·25 times as much power to fly at the same speed. Since the weight of flight muscle is more or less the same fraction of the total weight for all flying vertebrates (Greenwalt 1962), this means that heavier birds will have to produce more power per unit muscle weight than lighter birds. Hence the metabolic rate will have to be higher.

The disadvantage of weight cannot be accommodated by a reduction in the flying speed since a certain minimum speed is essential to produce the necessary lift. Furthermore, since larger birds have longer wing-spans, this leads to a reduction in the flapping frequency that can be maintained. If follows that there must be a limit to the weight of a bird dependent of flapping flight; according to Pennycuick (1972), this is about 12 kg with empty crop. The above of course, only applies to level flight; correspondingly more power will be expended in take off, but none at all in gliding or soaring.

Gliding and soaring

Several species, besides insects and birds have evolved a wing-like structure enabling them to glide, even if they cannot fly. The crudest structure is seen in the large, spreading webbed feet of the flying frog. The gliding lizard (*Draco*) has a fold of skin supported by its elongated ribs, whilst in the flying squirrel, this is stretched between the fore and hind limbs. In the flying fish, the pectoral fins are extended into wings when it leaps from the water.

Fig. 6.16 illustrates the situation for an animal gliding with steady speed U at an angle α to the horizontal. The forces on the body are W, the weight, acting vertically downwards, F_L the component lift-force generated by the wings and acting perpendicular to their surfaces, and F_D the drag force acting in line with the glide path.

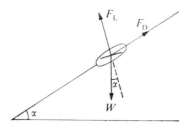

FIG. 6.16. Forces generated during gliding flight. For details see text.

In steady flight these forces must balance so that if we take the components parallel to and at right angles to the glide path, we have:

$$F_L = W \cos \alpha \quad \text{and} \quad F_D = W \sin \alpha,$$

whence,

$$\tan \alpha = \frac{F_D}{F_L} = \frac{C_D}{C_L}, \tag{6.11}$$

where C_D, C_L are the corresponding drag and lift coefficients.

It is evident from eqn (6.11) that the glide angle and therefore the distance covered by the glide, are independent of the weight; this distance is a maximum when tan α is a minimum i.e. when the lift/drag ratio, C_L/C_D is a maximum. As shown in Fig. 6.13 the lift/drag ratio depends on the shape and size of the wings and the angle of attack.

If the glide angle is small so that cos $\alpha \approx 1$, the lift can be assumed equal to the weight, so that eqn (6.9) also applies, i.e. $U = \sqrt{(2W/\rho A C_L)}$. For a given lift coefficient therefore, the speed of the glide increases with the wing loading factor. This can readily be confirmed by fastening a weight to a paper dart.

Increasing the angle of attack will increase the lift coefficient C_L and therefore reduce the glide speed, but as can be seen from Fig. 6.13, above a certain angle of attack, C_L falls very rapidly and at the stalling angle the lift is insufficient to balance the weight. Hence there will be a minimum speed at which a glide can be maintained.

The rate of descent, or sinking speed U_s, is given by the vertical component of the gliding speed, i.e. $U_s = U \sin \alpha$. For small values of α, it can be shown that:

$$U_s = \left(\frac{W}{\frac{1}{2}\rho A}\right)\left(\frac{C_D}{C_L^{\frac{3}{2}}}\right). \tag{6.12}$$

In other words, the sinking speed increases directly with the weight or wing loading factor, and inversely as the air density ρ.

The above relationship is important in thermal soaring, i.e. the ability of an animal to take advantage of rising thermal air currents, as in the case of man-made gliders. An animal will gain height as long as its sinking speed is less than that of the upward speed of the air current. Rooks, gulls, buzzards, and albatrosses are noted for their soaring abilities; all are large birds which experience some difficulty in normal flight because of their weight.

Albatrosses have aroused particular interest because of their ability to soar even when the wind movement is horizontal and to remain airborne for considerable periods without apparently moving their wings. As illustrated in Fig. 6.17, height is gained by gliding at speed *against* the wind, taking advantage of the increasing wind velocity with height above ground, and therefore, the increasing lift (A–C). Having gained a certain height, the bird turns and glides steeply downwind to gain speed (C–H), then it turns upwind at low level (H) to make a long, practically horizontal glide (H–J) before rising again. This long glide upwind makes up for ground lost in the preceding ascent, so enabling the bird to progress against the wind.

Turning (wheeling) and other manoeuvres are achieved by altering the position of the wings, either by rotating the body sideways (yaw) or lengthways (pitch). When the body is horizontal, the forces on the two wings are

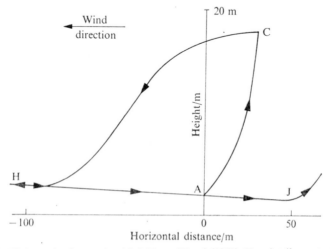

FIG. 6.17. Flight path of a soaring bird. From Wood (1973). For details see text.

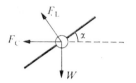

FIG. 6.18. Forces generated during wheeling. For details see text.

the same, but when turning, the outer wing moves faster and therefore generates a greater lift, causing it to rise above the other. As shown in Fig. 6.18 the resultant of the weight W acting vertically downwards, and the lift force F_L, now acting at an angle α to the vertical, is a force F_c acting inwards to the turn. If the turn is to be made without side slip, then:

$$F_c = \frac{WU^2}{r},\qquad(6.13)$$

where r is the radius of the turn.

A gliding bird achieves this by rolling its body through an angle α so that:

$$\tan\alpha = \frac{F_c}{W} = \frac{WU^2}{Wr} = \frac{U^2}{r}.\qquad(6.14)$$

7. Heat Transfer

THE temperature of an organism is determined by the balance between the heat produced by metabolic reactions within the body, e.g. by the oxidation of food or by muscular activity, and that arising from the exchange of heat between the body and its environment. Leaving aside heat losses due to evaporation, the transfer of heat within the systems concerned is brought about by three basic mechanisms, *conduction*, *convection*, and *radiation*. This chapter will be concerned only with conduction and convection; radiation will be discussed in Chapter 9 in relation to the general problem of heat balances.

Conduction

Heat conduction is essentially a molecular phenomenon. In solids and liquids it is generally believed that heat energy is transmitted, in part at least, by successive collisions between molecules. In gases, heat conduction arises primarily from the random movement of molecules, as has already been described for momentum transfer. Molecules with higher thermal energy moving randomly among molecules with lower thermal energy, collide and give up their energy as heat. The net effect is the transfer of heat from warmer to colder regions within a system.

Consider an element of conducting medium of area A and thickness δx with a steady temperature difference $\delta \theta$ between the two faces (Fig. 7.1). The rate of heat flow q_h (J s^{-1}) by conduction across the area will be proportional to A and $\delta \theta$, and inversely proportional to δx. For heat flow in one direction x, this relationship is formally expressed by Fourier's equation:

$$q_h = -\lambda A \frac{\delta \theta}{\delta x}.$$

In the limit, when $\delta x \to 0$, this becomes:

$$q_h = -\lambda A \frac{d\theta}{dx}, \tag{7.1}$$

(negative because heat moves in the direction of decreasing temperature).

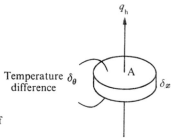

Temperature δ_θ
difference

FIG. 7.1. Heat conduction through a uniform slab of
medium (see text p. 80).

The flux of heat ϕ_h, i.e. the amount of heat crossing unit area of medium normal to the x direction in unit time, is then given by:

$$\phi_h = \frac{q_h}{A} = -\lambda \frac{d\theta}{dx}. \tag{7.2}$$

Similar equations can be written for heat conduction along other coordinates so that strictly speaking, the temperature gradients should be expressed in terms of partial differentials. For the sake of simplicity, we shall only consider conduction along the x-coordinate.

The proportionality factor λ represents the *thermal conductivity* of the medium and has units of J m^{-1} s^{-1} K^{-1} or W m^{-1} K^{-1}.

For conduction across a finite thickness of medium l, the heat flux through it is obtained by integrating eqn (7.2). Assuming that λ is constant,

$$\int_0^l \phi_h \, dx = -\lambda \int_{\theta_1}^{\theta_2} d\theta. \tag{7.3}$$

If the temperature difference and the heat flux ϕ_h are constant, this reduces to the well known equation for conduction in the steady state:

$$\phi_h = \lambda \frac{\theta_1 - \theta_2}{l} = \lambda \frac{\Delta\theta}{l}, \tag{7.4}$$

where θ_1, θ_2 correspond to the temperatures at the end faces ($\theta_1 > \theta_2$) and $\Delta\theta$ to $\theta_1 - \theta_2$. Thus, for a sheet of tissue of thickness $l = 1 \cdot 5$ mm, with a thermal conductivity $\lambda = 0 \cdot 3$ W m^{-1} K^{-1} and a temperature difference of $5\,°C$ across it, the flux of heat through it in the steady state will be $0 \cdot 3 \times 5 / 1 \cdot 5 \times 10^{-3} = 10^3$ J m^{-2} s^{-1}.

Conduction plays a major role in the transfer of heat between the surface of an organism and the internal tissues. Since most tissues consist largely of water, and water has a rather low thermal conductivity it would be expected that their thermal conductivities will also be low; in fact, as shown in Table 7.1, they are generally lower than that of water. However, as we shall see later, heat transfer through tissues provided with a blood supply may be

TABLE 7.1
Thermal conductivities of natural materials

	W m⁻¹ K⁻¹		
Human skin		Wood	
epidermis	0·21	parallel to axis	0·11 to 0·52
dermis	0·29 to 0·38	perpendicular to axis	0·04 to 0·20
fatty tissue	0·17 to 0·21	cell walls, parallel to fibres	0·65
muscle tissue	0·20 to 0·46	perpendicular to fibres	0·42
Rabbit fur	$3·77 \times 10^{-2}$	sapwood, heartwood	0·38, 0·15
Sheep wool	$4·19 \times 10^{-2}$		
Air (20 °C)	$2·57 \times 10^{-2}$	Water (20 °C)	0·60

Data from Tregear (1966), Siau (1971), Herrington (1966).

considerably enhanced by blood flow. Air has an even lower thermal conductivity than water. We see in Table 7.1 that when water is replaced by air, such as during the change from sapwood to heartwood in trees, there is a distinct fall in thermal conductivity. Several natural materials are distinctly anisotropic in their thermal properties, e.g. the cell wall material of wood has a much higher conductivity parallel to, than across the fibres, and this is reflected in the higher conductivity parallel to the stem axis, about $2\frac{1}{2}$ times that in the transverse direction when there is no free water present in the cells. Entrapped, still air is, or course, largely responsible for the excellent thermal insulation of animal pelts (see Table 7.1).

When it is required to calculate heat conduction through a number of tissues in series, each with a different thermal conductivity and thickness, the problem is simplified by introducing an insulation or thermal resistance factor I defined by:

$$\phi_h = \frac{\Delta\theta}{I}. \tag{7.5}$$

It follows from eqn (7.4) that $I = l/\lambda$, and has units of $J^{-1} m^2 s K$.

Eqn (7.5) is analogous to Ohm's law in the sense that heat flow is induced by a temperature difference whilst an electric current is induced by a voltage difference. The thermal resistance is therefore analogous to the electric resistance and just as we can add up electrical resistances in series so we can add up thermal resistances to obtain an overall resistance to the flow of heat.

Suppose, as illustrated in Fig. 7.2, we have three layers of tissue with thicknesses, l_1, l_2, and l_3, and thermal conductivities λ_1, λ_2, and λ_3; the corresponding thermal resistances will be l_1/λ_1, l_2/λ_2, and l_3/λ_3. If the overall temperature difference is $\Delta\theta$, the flux of heat ϕ_h through the three tissues in series will then be:

$$\phi_h = \frac{\Delta\theta}{l_1/\lambda_1 + l_2/\lambda_2 + l_3/\lambda_3}. \tag{7.6}$$

In certain cases it may be necessary to calculate the heat flow across the

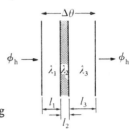

F<small>IG</small>. 7.2. Heat conduction through a composite slab varying
in thickness and thermal conductivity (see text p. 82).

wall of a hollow cylinder or sphere. Details are given in Appendix 12. For
a cylinder of length l, with external and internal radii, r_o, r_i respectively,
the solution is:

$$q_h = \frac{2\pi\lambda l(\theta_o - \theta_i)}{\ln(r_o/r_i)} \qquad (7.7)$$

where q_h is the rate of heat flow through the whole wall (J s^{-1}), λ the thermal
conductivity of the wall material (assumed constant) and θ_o, θ_i, respectively,
the temperature of the outer and inner surfaces; it is assumed that $\theta_o > \theta_i$
i.e. heat travels from the outer to inner surface, but the equation can be
easily modified for flow in the other direction.

It follows from eqn (7.7) that the rate of heat conduction across the walls of
a cylinder depends on the ratio of the external and internal radii, r_o/r_i, not
on their absolute values. In the case of very thin walled tubes, when $r_o - r_i$ is
very small, heat conduction can be estimated approximately by regarding the
cylinder as opened out into a slab of uniform thickness $r_o - r_i$, with an area
equal to the arithmetic mean of the inner and outer surfaces.

The flux of heat, i.e. the rate of heat flow through unit area of wall will
differ for the two surfaces because of their different areas. For any one surface
it can be readily obtained by dividing q_h in eqn (7.7) by $2\pi r l$. For the outer
surface with radius r_o therefore, we have:

$$\phi_{h(o)} = \frac{\lambda(\theta_o - \theta_i)}{r_o \ln(r_o/r_i)}. \qquad (7.8)$$

For a hollow sphere, with internal and external radii r_i, r_o, the rate of
heat flow through the whole wall is given by:

$$q_h = \frac{4\pi(\theta_o - \theta_i)}{(1/r_i - 1/r_o)}, \qquad (7.9)$$

and the flux of heat through say, the outer wall will be:

$$\phi_{h(o)} = \frac{\lambda(\theta_o - \theta_i)}{r_o^2(1/r_i - 1/r_o)}. \qquad (7.10)$$

We see that with a hollow sphere, the rate of conduction is not independent
of the absolute size.

7

If the walls consist of different layers differing in thickness and thermal conductivity, as for example in blood vessels, the solution is again most readily obtained on the basis of the thermal resistance of each layer.

Non-steady heat flow

So far we have dealt with heat conduction in the steady state, i.e. with a constant heat flux through the medium induced by a constant temperature difference across it. Under these circumstances, the temperature at any point within the medium remains constant with time. There are however, many cases of biological interest in which the medium absorbs heat so that, as heat flows through it, the temperature at a given point changes with time. It can then be shown (Appendix 13) that the rate at which the temperature of the medium rises $(d\theta/dt)$ is given by:

$$\frac{d\theta}{dt} = \frac{\lambda}{\rho c_p} \cdot \frac{d^2\theta}{dx^2} = \alpha \frac{d^2\theta}{dx^2}, \tag{7.11}$$

where λ, ρ, and c_p are respectively the thermal conductivity, density and specific heat of the medium.

In other words, the rate at which the temperature rises is proportional to the rate at which the temperature gradient $(d\theta/dx)$ changes with distance x along the direction of heat flow. The proportionality constant α $(= \lambda/\rho c_p)$ is known as the *thermal diffusivity* and has units of $m^2\ s^{-1}$, the same as those of the kinematic viscosity, ν.

Soil temperature fluctuations

There are several possible solutions to eqn (7.11) and the choice depends on the particular problem to be solved (Carslaw and Jaeger 1959). A commonly quoted problem which is of some biological interest is how the temperature of a soil changes with depth as a result of temperature fluctuations at the surface, arising either diurnally (day–night) or annually (summer–winter).

Suppose that the surface temperature (θ_s) alternates sinusoidally above and below a mean temperature (θ_a) with an amplitude A (K) and a frequency of f cycles per day (this is a more realistic time basis than the conventional second since f is now 1 for the diurnal, and $\frac{1}{365}$ for the annual cycle (Fig. 7.3).

If, for the sake of simplicity, $\theta_s = \theta_a$ at the beginning of a given period (time $t = 0$), then after an interval of time t (days) the surface temperature is given by:

$$\theta_s = \theta_a + A \sin(2\pi ft), \tag{7.12}$$

where $2\pi f$ corresponds to the angular velocity in radians per day (Fig. 7.3).

Thus, for the diurnal cycle with $f = 1$, the surface temperature after say, 6 hours ($t = \frac{1}{4}$ day) will be $\theta_a + A \sin \pi/2 = \theta_a + A$.

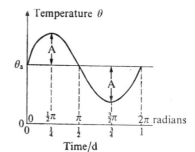

FIG. 7.3. Diurnal variation in soil surface temperature, assumed sinusoidal, with equivalent time scale in radians.

For a semi-infinite medium with uniform thermal diffusivity α, the appropriate solution to eqn (7.11) which gives the temperature $\theta_{z,t}$ at a depth z in the soil at time t is:

$$\theta_{z,t} = \theta_a + Ae^{-cz}\sin(2\pi ft - cz) \quad \text{where} \quad c = \sqrt{\left(\frac{\pi f}{\alpha}\right)}. \quad (7.13)$$

At the surface ($z = 0$) this equation reverts to eqn (7.12).

Several points of interest arise from eqn (7.13) and as a check, some experimental data for diurnal soil temperatures at different times of the day and at different depths are illustrated in Fig. 7.4. The time scale will have to be adjusted to fit the assumption that $\theta_s = \theta_a$ at time $t = 0$.

It follows from eqn (7.13) that at a depth z the temperature varies with an amplitude Ae^{-cz} so that as z increases, the amplitude decreases; in other words, as is clearly shown in Fig. 7.4, the greater the depth, the less the fluctuation in temperature.

The amplitude also decreases as c increases, i.e. as $\sqrt{f}$; the more rapid the surface fluctuation of temperature therefore, the less will the temperature vary at a given depth. Since the frequency of the diurnal fluctuation is 365 times greater than that of the annual fluctuation, the variation in temperature will be $\sqrt{365}$ or about 19 times less at the same depth. Measurements have shown that whereas diurnal fluctuations of temperature at the surface have relatively little effect on temperatures at depths greater than about 1 m (compare Fig. 7.4), seasonal fluctuations may penetrate as deep as 20 m or more. Clearly, this will have important consequences concerning the biology of soil organisms and plant roots.

The time lag between minimum or maximum temperatures at the surface and the corresponding minima or maxima at a depth z is given by cz. This means that the lag should increase linearly with depth (as suggested by the dashed line in Fig. 7.4), and with the square root of the frequency. The lag should therefore be much greater for diurnal than for annual surface temperature fluctuations; according to Fig. 7.4, it is of the order of 3 hours, even at a depth of only 0·1 m.

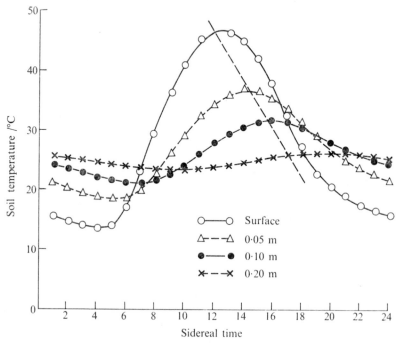

FIG. 7.4. Diurnal variation in soil temperature with depth. After Yakuwa (1945).

It is evident from eqn (7.13) that the thermal diffusivity of the soil (α) is an important factor determining the amplitude and lag of the temperature fluctuation. In turn α depends on the relative proportions of solid, liquid, and gas phases in the soil, and the size and arrangement of the particles. The data summarized in Table 7.2 illustrate the profound effect of water content on the soil thermal diffusivity, especially for dry mineral soils.

TABLE 7.2
Thermal properties of soils

	Water content volume %	Density kg m^{-3}	Specific heat 10^3 J kg^{-1} K^{-1}	Thermal conductivity W m^{-1} K^{-1}	Thermal diffusivity 10^{-6} m^2 s^{-1}
Sand	0	1·60	0·80	0·30	0·24
(40% pore space)	20	1·80	1·18	1·80	0·85
	40	2·00	1·48	2·20	0·74
Clay	0	1·60	0·89	0·25	0·18
(40% pore space)	20	1·80	1·25	1·18	0·53
	40	2·00	1·55	1·58	0·51
Peat	0	0·30	1·92	0·06	0·10
(40% pore space)	20	0·70	3·30	0·29	0·13
	40	1·10	3·65	0·50	0·12

Data from Monteith (1973), van Wijk and de Vries (1963).

A somewhat similar problem is provided by the temperature distribution in tree stems as a result of diurnal fluctuations in the ambient air temperature (Herrington 1969).

Convection

Most readers will be familiar with the general principles of convection. If say, a beaker of water is heated from below, then since water has a very low thermal conductivity, the layer at the bottom will rise in temperature much faster than the bulk water above. As a result of its volume expansion and corresponding decrease in density, this warmer layer of water will rise through the bulk water and will be replaced by colder water, thus setting up a circulatory current in which heat is transported 'bodily' by the warmer water. This will be maintained as long as there is a temperature gradient within the water. Exactly the same happens in air above the ground if the latter is at a higher temperature.

The transport of heat by a moving fluid is known as *convection* and is very much more rapid than that by conduction. *Free* or *natural convection* is said to occur if, as in the water example above, fluid movement is induced solely by temperature and density changes. If the fluid is moved by an external force, e.g. by stirring the water in the beaker, or in the case of air, by wind, the process of heat transport is then called *forced convection*.

The freezing of ponds

The way in which a pond freezes provides a fairly simple environmental example of free convection combined with two further facts: (1) water has has a maximum density at about 4 °C and (2) heat is lost only from the upper surface.

Suppose that the air above the surface is at say, -10 °C. Heat will be lost from the surface layer of water to the air, the density of this water will rise and as it sinks towards the bottom of the pond, it will be replaced by warmer and therefore less dense water from below. Free convection will continue until the whole of the water in the pond is at 4 °C.

As more heat is lost to the air, the surface water will continue to cool down, but now being less dense, it will remain at the surface so that eventually a thin layer of ice will form.

To cool 1 kg liquid water from 4 °C to 0 °C requires the abstraction of $1 \cdot 68 \times 10^4$ J; the latent heat of fusion of ice (L_f) is $3 \cdot 35 \times 10^5$ J kg^{-1} so this represents the extra heat that must be abstracted to convert each kg liquid water at 0 °C into ice at 0 °C, giving a total heat abstraction of about $3 \cdot 52 \times 10^5$ J kg^{-1}. The rate at which the ice layer thickens therefore, will be determined by the rate at which this amount of heat is conducted from the water below to the air above.

Strictly speaking, the solution to this problem should take into account conduction through both water and ice, but we shall simplify it by considering only conduction through the ice. It can then be shown (Appendix 14) that the rate at which the ice layer thickens at any given time is inversely proportional to the existing thickness. If ρ_I = density of the ice at 0 °C (kg m^{-3}) and its thermal conductivity is λ (W m^{-1} K^{-1}),† then assuming that the lower surface of the ice is at 0 °C and its upper surface at $-\theta$ °C (the air temperature), we have:

$$l^2 = \frac{2\lambda\theta t}{L_t \rho_I} \tag{7.14}$$

where l = thickness of the ice at time t(s) and L_t = latent heat of fusion. From this we see that the thickness of the ice increases as the square root of the time.

Thermal boundary layers

Consider a vertical plate in contact with a fluid at a lower temperature (Fig. 7.5). As the fluid next to the plate is warmed and its density falls, it will rise past the plate to produce a boundary layer in which, depending on the speed of the fluid, its viscosity, and the length of surface, flow may be laminar or turbulent. This we call the *hydrodynamic boundary layer*. In the case of free convection (Fig. 7.5a), the velocity of flow parallel to the surface

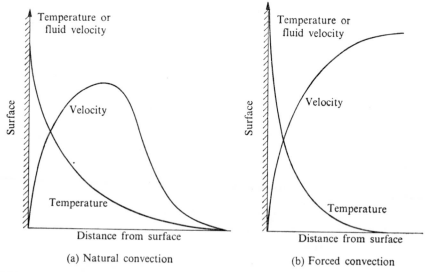

(a) Natural convection   (b) Forced convection

FIG. 7.5. Theoretical velocity and temperature profiles for (a) free and (b) forced convection from a vertical plane surface.

† As a unit of temperature interval, 1°C $\equiv$ 1K.

will increase from zero at the surface itself to a maximum at some distance from the surface and will then decrease again to zero corresponding to the stationary bulk fluid. With forced convection (Fig. 7.5b), when the fluid is impelled past the plate by an external force, the fluid velocity will increase with distance from the surface, gradually approaching that of the bulk fluid (as Fig. 4.1).

With both modes of convection, the temperature decreases with distance from the surface until it attains that of the bulk fluid (Fig. 7.5a,b). By analogy with the hydrodynamic boundary layer, the region near the surface over which the fluid temperature changes is known as the *thermal boundary layer*. The effect of convection therefore, is to produce a temperature gradient within the fluid normal to the surface of the plate, and the rate at which heat travels down this gradient across the thermal boundary layer provides us with a measure of the rate of convection from surface to fluid. Now when flow in the surface boundary layer is laminar, the only way in which heat can cross this layer is by molecular diffusion, i.e. by conduction. The same also applies, though to a much smaller extent, when flow is turbulent because of the laminar sub-layer next to the surface (p. 41); we assume that mixing is so thorough within the turbulent boundary layer proper that there is no longer any appreciable variation in temperature, i.e. that the temperature of the bulk fluid is uniform.

For a given temperature difference between the surface and bulk fluid, the temperature gradient will be determined by the depth of the thermal boundary layer; the thinner this layer, the steeper will be the gradient and as is evident from the conduction eqn (7.1), the more rapid will be the rate of heat transfer. Clearly a knowledge of how the depth of this thermal boundary layer varies with flow conditions is fundamental to our understanding of the factors determining convection rates. In this respect a very useful relationship has been established between the depth of the thermal boundary layer (δ_{th}) and that of the hydrodynamic boundary layer (δ_b). On the argument that δ_{th} will be related to the thermal diffusivity of the fluid (α) in the same way as δ_b is related to its kinematic viscosity (ν), it has been shown that for most fluids, including air and water:

$$\frac{\delta_b}{\delta_{th}} = \left(\frac{\nu}{\alpha}\right)^{\frac{1}{3}}. \tag{7.15}$$

Since ν and α have the same dimensions (m^2 s^{-1}), the above ratio is dimensionless and is known as the Prandtl number (*Pr*). We shall see later that this is a very important quantity governing convective heat transfer.

As we are already familiar with the factors determining δ_b (see Chapter 4) we might anticipate some of the factors determining δ_{th}. For example, in the case of flow over a vertical plane surface, δ_b increases with distance from the leading edge (eqns 4.1 and 4.5) so δ_{th} will also increase; this means that the local rate of convection will decrease. In other words, with a uniform temperature difference between surface and bulk fluid, the rate of heat transfer will vary at different points along the surface. As in the case of friction drag therefore, (p. 41) we must distinguish between local rates of convection and overall, or average rates for the surface as a whole. Another conclusion to be drawn from the relationship between δ_b and δ_{th} is that at any given point on a surface, δ_{th} will decrease with increasing velocity of flow past the surface, so steepening the temperature gradient. This is clearly shown in Fig. 7.6 for the

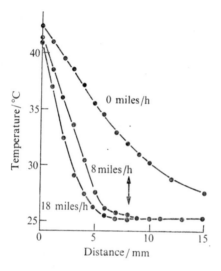

FIG. 7.6. Variation in air temperature (°C) with distance from the body surface of an animal at different windspeeds (1 mile h^{-1} = 0·45 m s^{-1}). From Tregear (1965).

effect of wind-speed over the body of an animal. It will be noted from this figure that like the hydrodynamic velocity gradient, the temperature change within the thermal boundary layer is not linear; hence besides δ_{th}, we would have to know how this gradient varies with distance from the surface before we can evaluate the rate of convective heat transfer.

It will be appreciated that the rate of convection will be governed by several factors. In practice, it is customary to by-pass the complex relationships that are involved by simply expressing the convective heat flux (ϕ_h) in terms of the temperature difference ($\Delta\theta$) between a surface and bulk fluid and an appropriate coefficient:

$$\phi_h = h\Delta\theta. \tag{7.16}$$

The coefficient h, variously known as the film, heat transfer or convection coefficient (units J m^{-2} s^{-1} K^{-1}) may refer to a local (h) or average for the whole surface ($\bar{h}$). It will vary according to the shape and orientation of the surface and flow conditions, and to a large extent, these factors determine how it can be evaluated. However, except for forced convection with laminar flow over a plane surface, which can be dealt with from first principles, all other solutions rest ultimately on experiment and measurement. The problem then remains as to how to deal with the several variables that are involved. Here we find another application of dimensional analysis.

Free convection for example, involves the transport of heat by a mass of fluid moving away from a surface by buoyancy forces due to changes in temperature and density. We might deduce therefore, that the following parameters will be involved: the temperature difference between the surface and bulk fluid ($\Delta\theta$); the density of the fluid (ρ), its viscosity (η or ν); specific heat (c_p); thermal conductivity (λ); and temperature coefficient of expansion (β)—also a dimension of length (l) characterising the surface and the acceleration due to gravity (g). It turns out that as in the case of the Reynolds' number, certain of these parameters can be grouped into dimensionless quantities each of which describes a particular aspect of fluid behaviour important in free convection. These quantities are:

Nusselt number, (Nu) $= hl/\lambda$

This relates the convection coefficient h to the thermal conductivity of the fluid (λ) and a length (l) corresponding to the length or diameter of the body. (Nu) can be considered as the ratio of the convective heat flux ($\phi_h = h\Delta\theta$) to the conductive heat flux ($= \lambda\Delta\theta/l$). As in the case of h and $\bar{h}$, we shall distinguish between (Nu) for local and ($\overline{Nu}$) for average heat transfer from a surface.

Prandtl number (Pr) $= \nu/\alpha = c_p\eta/\lambda$

As we have already seen (eqn 7.15), this is equivalent to the ratio of the kinematic viscosity of the fluid ($\nu = \eta/\rho$) to its thermal diffusivity ($\alpha = \lambda/\rho c_p$), and relates the depths of the hydrodynamic and thermal boundary layers.

Grashof number (Gr) $= \dfrac{g\beta l^3\Delta\theta}{\nu^2}$

This quantity is concerned specifically with free convection and represents the ratio of buoyancy forces on the fluid (g, β, $\Delta\theta$) to viscous drag forces ($\eta = \rho\nu$). As before, l refers to the length or diameter of the body.

When theoretical or experimental results for heat transfer are expressed in terms of these three dimensionless numbers, it is found that with free convection, a consistent general relationship can be established between them over a wide range of flow conditions. Fig. 7.7 illustrates the results of a number of experiments on free convection in which ($\overline{Nu}$) is plotted against ($Gr.Pr$) on a log–log scale. We see that over the range ($Gr.Pr$) $= 10^3$ to 10^8, corresponding to laminar flow, a reasonably good fit to the curve is given by the empirical equation:

$$(\overline{Nu}) = 0.56(Gr.Pr)^{\frac{1}{4}}, \qquad (7.17)$$

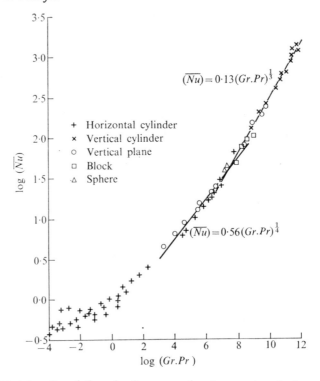

FIG. 7.7. Heat transfer relations for free convection from various body surfaces with equations fitted to the curve. After Jakob (1949).

whilst for higher $(Gr.Pr)$ values, when a turbulent boundary layer is formed, the curve can be fitted to the equation:

$$(\overline{Nu}) = 0.13(Gr.Pr)^{\frac{1}{3}}. \tag{7.18}$$

Although more precise relationships may be derived for certain systems, in most cases of free convection, (Nu) (or $\overline{Nu}$) can be expressed with reasonable accuracy as a function of $(Gr.Pr)^n$ where $n = \frac{1}{4}$ for laminar flow and $\frac{1}{3}$ for turbulent flow.

A similar approach may be adopted for forced convection and it is then found that in most cases (Nu) (or $\overline{Nu}$) can be expressed as a function of (Pr) and the Reynold's number, (Re) with the general form:

$$(Nu) = c(Pr)^m(Re)^n \tag{7.19}$$

where c, m, and n have values determined by the shape of the surface and flow conditions.

Once (Nu) or $(\overline{Nu})$ is known for a given system, h or $\bar{h}$ can be calculated from $(Nu) = hl/\lambda$ for a given fluid and body dimensions, and then, for a known temperature difference ($\Delta\theta$) the rate of convective heat transfer per unit surface area (ϕ_h) can be calculated from eqn (7.16). (see below).

Equations for $(\overline{Nu})$ for free and forced convection from smooth plates, cylinders and spheres under different flow conditions are summarized in Tables 7.3 and 7.4; solutions for the heat flux ϕ_h in air at 20 °C calculated from $\phi_h = \{(\overline{Nu})\lambda/l\}\Delta\theta$ are given in Table 7.4 for forced convection. The following points should be borne in mind:

1. In all cases the relevant physical properties of the fluid (ν, λ etc.) refer to those at the arithmetic average temperature of the surface and bulk fluid. To ease computation, values for $(Gr)/l^3\Delta\theta$ and $(Pr.Gr)/l^3\Delta\theta$ are included in Table 1.1 for water and air at different temperatures.

2. Strictly speaking, the given formulae only apply within the conditions under which they were obtained; i.e. within the stated limits of $(Gr.Pr)$ or (Re). Reference to the literature will disclose some slight variation in the constants depending on the author's choice of experimental results; this reflects the empirical nature of the equations. Most of the solutions given apply to air and water, i.e. to a fairly wide range of (Pr); some however, are restricted to $(Pr) \approx 1$ and apply only to air and other gases.

3. Except for horizontal plane surfaces with free convection, the same equation applies to surfaces in contact with warmer fluids.

4. When the surface is held at an increasing angle to the fluid stream, heat transfer by forced convection is increased; data for a cylinder are illustrated in Fig. 7.8. It will be noted that when in line with the flow, the convection rate is only about 40% of that at right angles to the flow.

5. All equations quoted refer to surfaces with a uniform temperature, such as might be assumed for metal surfaces with relatively high thermal conductivities. Most natural surfaces however, have relatively low conductivities so that the temperature may vary appreciably over the surface, depending on local convection rates. This problem, together with others associated with living organisms will be discussed later.

6. It will be noted that for free convection with turbulent flow, $(\overline{Nu}) \propto (Gr)^{\frac{1}{3}}$ i.e. $(\overline{Nu}) \propto l$; this means that h which is proportional to $(\overline{Nu})/l$ will be independent of body size. However, there are probably few cases in nature when a turbulent boundary layer is formed by free convection. With laminar flow, $(\overline{Nu}) \propto (Gr)^{\frac{1}{4}}$ whence $h \propto (\Delta\theta)^{\frac{1}{4}}$ and $\phi_h \propto (\Delta\theta)^{\frac{5}{4}}$; this corresponds to the well known $\frac{5}{4}$ power law for cooling by free convection.

7. Although a distinction has been made between free and forced convection, the two processes cannot be separated so precisely and in many cases, both contribute to heat transfer (so-called *mixed convection*). Their relative contributions may be assessed by comparing (Gr) with

TABLE 7.3
Free convection

	Range	Flow (fluid)	$(\overline{Nu})$
Vertical plane surfaces, vertical, large diameter cylinders	$10^4 < (Gr.Pr) < 10^9$ $(Gr.Pr) > 10^9$ $(Gr.Pr) > 2 \times 10^9$	Laminar (a, w) Turbulent (a) Turbulent (w)	$0.56(Gr.Pr)^{\frac14}$ $0.13(Gr.Pr)^{\frac13}$ $0.17(Gr.Pr)^{\frac13}$
Horizontal plane, warm face up or cold face down	$(Gr) < 10^5$ $(Gr) > 10^5$	Laminar (a) Turbulent (a)	$0.54(Gr.Pr)^{\frac14}$ $0.14(Gr.Pr)^{\frac13}$
Horizontal plane; cold face up or warm face down	$(Gr.Pr) < 10^{10}$	Laminar (a) No turbulence	$0.25(Gr.Pr)^{\frac14}$
Long horizontal or vertical cylinders	$10 < (Gr.Pr) < 10^3$ $10^3 < (Gr.Pr) < 10^8$ $(Gr.Pr) > 10^8$	Laminar (a, w) Laminar (a, w) Turbulent (a, w)	$0.91(Gr.Pr)^{0.17}$ $0.47(Gr.Pr)^{\frac14}$ $0.10(Gr.Pr)^{\frac13}$
Spheres	$(Gr)^{\frac14} < 220$	(a)	$2 + 0.54(Gr)^{\frac14}$

l = mean side

Fluid, *a* = air, w = water

Data from Fishenden and Saunders (1950), Bird, Stewart and Lightfoot (1960), Ede (1967).

TABLE 7.4

Forced convection

	Flow (fluid)	Range	$(\overline{Nu})$
Plane surfaces	Laminar (a, w)	$(Re) < 2\times10^4$	$0{\cdot}664(Pr)^{\frac{1}{3}}(Re)^{\frac{1}{2}}$
	Laminar (a)		$0{\cdot}60(Re)^{\frac{1}{2}}$
	Turbulent (a, w)	$(Re) > 2\times10^4$	$0{\cdot}036(Pr)^{\frac{1}{3}}(Re)^{0.8}$
	Turbulent (a)		$0{\cdot}032(Re)^{0.8}$

	Flow (fluid)	Range	$(\overline{Nu})$
Cylinders	(a, w)		$c(Re)^m$

(Re)	c	m
1–4	0·891	0·330
4–40	0·821	0·385
40–4×10^3	0·615	0·466
4×10^3–4×10^4	0·174	0·618
4×10^4–4×10^5	0·024	0·805

	Flow (fluid)	Range	$(\overline{Nu})$
	(a)	$10^{-1} < (Re) < 10^3$	$0{\cdot}32+0{\cdot}51(Re)^{0.52}$
	(a)	$10^3 < (Re) < 10^5$	$0{\cdot}24(Re)^{0.60}$
Spheres	(a)	$0 < (Re) < 300$	$2+0{\cdot}54(Re)^{\frac{1}{2}}$
	(a)	$50 < (Re) < 1{\cdot}5\times10^5$	$0{\cdot}34(Re)^{0.6}$

Heat flux $(\phi_h\ W\ m^{-2})$ *in air at* $20°C$ *for temperature difference* $\Delta\theta(°C)$

$(\lambda = 2{\cdot}57\times10^{-2}\ W\ m^{-1\circ}C^{-1}; \nu = 1{\cdot}51\times10^{-5}\ m^2\ s^{-1}; U = $ windspeed, $m\ s^{-1})$

(length of surface l or body diameter d in m)

Plane surfaces: laminar flow $3{\cdot}97\sqrt{(U/l)}\Delta\theta$

                 turbulent flow $5{\cdot}91(U^{0.8}/l^{0.2})\Delta\theta$

Cylinders: $10^3 < (Re) < 10^5$ $4{\cdot}82(U^{0.6}/l^{0.4})\Delta\theta$

Spheres: $50 < (Re) < 1{\cdot}5\times10^5$ $6{\cdot}82(U^{0.6}/d^{0.4})\Delta\theta$

Fluid, a = air, w = water

Data of Fishenden and Saunders (1950), Bird, Stewart, and Lightfoot (1960), Ede (1967).

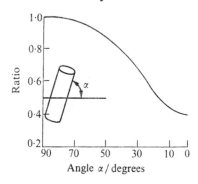

FIG. 7.8. Forced convection coefficient for a cylinder at different angles (α) to the direction of air flow, expressed as the ratio of the convection coefficient at $\alpha = 90°$.

$(Re)^2$. Since (Gr) represents the ratio of buoyancy forces to (viscous force)^2 and (Re), the ratio of inertial to viscous forces, the ratio $(Gr)/(Re)^2$ represents the ratio of buoyancy to (inertial forces)^2, i.e. of free to forced convection. When this ratio is high (above about 16) free convection predominates, when it is low (below about 0·1), forced convection predominates. Between these limits mixed convection occurs and it is then customary to calculate $\bar{h}$ for both and choose the higher value.

As an example, consider the case of a vertical square plate of side 0·1 m with a uniform surface temperature of 22·5 °C, over which air is blown parallel to one side at a uniform speed of U m s^{-1}. If the bulk temperature of the air is 17·5 °C, the temperature difference ($\Delta\theta$) will be 5 °C and the average temperature of the system, 20 °C.

According to Table 1.1, for air at 20 °C with $l = 0·1$ and $\Delta\theta = 5$,

$$(Gr) = 1·47 \times 10^8 \times (0·1)^3 \times 5 = 7·35 \times 10^5.$$

At the same 20 °C, ν for air $= 1·5 \times 10^{-5}$ m^2 s^{-1}, so that

$$(Re) = U \times 0·1/1·5 \times 10^{-5} = 6·7 \times 10^3 \ U,$$

and

$$(Re)^2 = 45 \times 10^6 \ U^2.$$

The ratio $(Gr)/(Re)^2$ then $= 1·6 \times 10^{-2}/U^2$.

When U exceeds 0·4 m s^{-1}, the ratio is less than 0·1 so that forced convection will predominate; when U is less than 0·03 m s^{-1}, the ratio exceeds 16 and free convection predominates.

If $U = 0·5$ m s^{-1}, $(Re) = 0·5 \times 0·1/1·5 \times 10^{-5} = 3·3 \times 10^3$. According to Table 7.4, flow should be laminar and $(\overline{Nu}) = 0·6(Re)^{\frac{1}{2}} = 34·5$.

For air at 20 °C, $\lambda = 2·57 \times 10^{-2}$ W m^{-1} s K^{-1}, whence

$$\bar{h} = \lambda(\overline{Nu})/l = 2·57 \times 10^{-2} \times (34·5/0·1) = 8·86 \text{ W m}^{-2} \text{ K}^{-1}.$$

From eqn (7.16), $\phi_h = \bar{h}\Delta\theta = 8.86 \times 5 = 44·3$ W m^{-2}. Since the total area of the plate (both sides) is 0·02 m^2, the rate of convective heat loss from the whole plate will be 0·89 W.

If $U = 0.1$ m s^{-1}, the ratio $(Gr)/(Re)^2$ lies between the stated limits so that mixed convection would be expected. Since $(Re) = 6.7 \times 10^2$, i.e. less than 2×10^4, flow will be laminar (Table 7.4). With *free convection*, $(\overline{Nu}) = 0.56$ $(Gr.Pr)^{\frac{1}{4}}$; at 20 °C, (Pr) for air $= 0.71$ and $(Gr) = 7.35 \times 10^5$ (as above); therefore, $(\overline{Nu}) = 15$. With *forced convection*, $(\overline{Nu}) = 0.6(Re)^{\frac{1}{2}} = 15.5$. In this case therefore, there is little difference between heat losses by free and forced convection.

It is of interest that for the same plate in an aqueous medium, with the same temperature regime, (Gr) is about 10 times higher, but $(Re)^2$ over 200 times higher than in air for the same fluid velocity. Since this gives a much lower $(Gr)/(Re)^2$ ratio, forced convection will predominate at much lower speeds.

Biological applications

A considerable number of measurements have been made of convective heat transfer from organisms, though almost exclusively in air and usually as part of a more extensive study on heat balances.

Leaves

Several studies on flat leaves have confirmed that even though they do not strictly conform to plane surfaces, the effect of curvature is usually small. Under normal wind conditions, (Re) for leaves is of the order of 10^2 to 10^4 so that laminar flow can usually be assumed (see also p. 98). According to Table 7.4 for forced convection with laminar flow over a plane surface, $(\overline{Nu})$ is proportional to $(Re)^{\frac{1}{2}}$ so that the average convection coefficient $\bar{h} \ (= (\lambda/l)(\overline{Nu}))$ will be proportional to $(Re)^{\frac{1}{2}}/l$ i.e. to $\sqrt{(U/l)}$, where U is the wind-speed. Most measurements on leaves have confirmed this relationship. It follows that the higher the wind velocity and/or the smaller the dimension of the leaf in the direction of wind movement, the greater will be the rate of heat transfer; in other words, the smaller the leaf, the more rapidly will it equilibrate with the ambient air temperature. This too is confirmed by experience and applies also to cylindrical shaped leaves like those of conifers, for which the appropriate dimension is the diameter.

There are however, certain features of leaves which may lead to quantitative relationships, very different from those given in Table 7.4 for forced convection from flat plates. From several measurements reported in the literature, Monteith (1965) concluded that in the open, $\bar{h} \approx 9\sqrt{(U/l)}$, i.e. about $2\frac{1}{2}$ times the theoretical value; (n.b. Monteith expresses the data in terms of a diffusion resistance to heat transfer, $r_{ah} = \rho c_p/\bar{h}$ (eqn 8.17 and p. 124). As already mentioned, a feature of poorly-conducting bodies like leaves is the non-uniform temperature distribution over the surface. According to Parlange, Waggoner, and Heichel (1971), the convective heat loss is more likely to be uniform over a leaf surface than the temperature, and under these circumstances the temperature increases as the square root of the distance from the leading edge. They showed that with steady laminar air flow, the convective heat loss would be similar to that from a surface with uniform temperature and they concluded that the higher heat losses from leaves in the open were due to wind turbulence. Measurements on a leaf placed in front of an electric fan with air speeds ranging from 0·6 to 1·6 m s^{-1} confirmed almost exactly the $2\frac{1}{2}$ times increase in convection reported by Monteith, irrespective of whether the leaf was stationary or allowed to flap. The authors suggest that the effect may be due to boundary layer separation; an increase in heat transfer due to turbulence in the free stream has also been observed with cylinders (Kutateladze 1963).

Another feature of leaves which complicates the application of conventional formulae established for flat plates is their variable shape. Parlange *et al.* (1971) describe a method for calculating the average width of a leaf, defined as the width of a rectangle with the same longer dimension and the same average heat flux due to forced convection. The resulting expressions differ according to whether the temperature or heat flux is constant over the surface, but unless the leaf is very irregular, both cases yield similar values for the average width.

A very different set of circumstances arises when the leaf is deeply lobed, as indeed is often the case. Vogel (1970) compared convective heat losses from thin metal plates of the same surface area but differing in shape, by measuring the amount of electrical heat required to maintain the plates at a temperature 15 °C above that of the ambient air with wind speeds ranging from 0 to 0·3 m s^{-1}. He showed that the more irregular the outline, the higher the rate of heat dissipation; the greatest difference between lobed and circular plates occurred when air flow was perpendicular to the surface. In still air however, when only free convection took place, circular plates dissipated about 25 per cent more heat when vertical than horizontal, whereas with extensively lobed models, heat dissipation was largely independent of orientation. Unfortunately, no entirely consistent relationship could be established between plate shape and heat loss to explain these findings. From an ecological point of view, it seems that to some extent at least, shape could be related to an adaptation of leaves in nature to reduce the difference between leaf and air temperature. As we have just seen this can also be achieved by reducing the size of the leaf.

Although convective losses from single leaves are of some interest, in the natural state we are concerned more with arrays of leaves, and in the case of trees or shrubs, with complete canopies. In Chapter 9 we shall discuss other approaches to the problem of convective heat losses from vegetative covers, but at least one direct measurement has been made on conifer shoots (Tibbals, Carr, Gates, and Kreith, 1964). Silver castings were made of a spruce and fir shoot and free and forced convection losses measured in a wind tunnel at different wind speeds and different temperatures. With natural convection, both models behaved more or less like horizontal cylinders. With forced convection, in line with and across the air stream, the corresponding convection coefficients for the fir model lay on either side of those for an array of cylinders; the coefficients for the spruce model were much lower and this was attributed to interference with the air flow by the numerous layers of densely-packed needles.

Still another feature common to many leaves, and also to other organisms, is their rough surface, whereas the formulae quoted in Table 7.4 apply to smooth surfaces. We shall see later that convective heat transfer is related to skin friction drag; for the present it may be anticipated that, like friction drag (p. 42), heat transport is unaffected by surface roughness when flow is laminar, and only increases with turbulent flow if the average height of the roughness elements is greater than the thickness of the laminar sub-layer. The presence of hair, etc. on the surface will often reduce convection losses quite substantially by reducing or eliminating air movement next to the surface.

Animals

The effect of body shape and size on convective cooling of insects has been studied, among others, by Church (1960) who used freshly-killed bodies artificially heated by high-frequency induction. On the argument that the power required for flight would be approximately proportional to body weight, the rate of heating was adjusted to be proportional to the volume of the insect, i.e. proportional to their diameter³ (d^3). For moths and bumble bees with average thoracic diameters of about 10 mm and body volumes of about 525 mm³, heating was at the rate of 190 m W. It was then assumed that the excess temperature of the body surface over that of the ambient air would be similar to that in normal flight.

From wind-tunnel measurements on insects denuded of body hair, with a wind speed of 3 m s⁻¹, it was found that the temperature excess within the pterothorax ($\Delta\theta_t$) increased as the 1·3 to 1·4 power of the average body diameter, i.e. $\Delta\theta_t \propto d^{(1\cdot3-1\cdot4)}$. Because of the low thermal conductivities of the body tissue the temperature excess at the surface of the body ($\Delta\theta_s$) was only about 10 per cent lower than that in the pterothorax and would still be proportional to $d^{(1\cdot3-1\cdot4)}$.

With a wind-speed of 3 m s⁻¹, (Re) for the different insects studied ranged from about 10 to 10^3; at these Reynolds' numbers, forced convection from cylinders and spheres is probably better represented by $(\overline{Nu}) \propto (Re)^{0\cdot5}$, but Church continues to use $(Re)^{0\cdot6}$. Substituting $(\overline{Nu}) = \bar{h}d/\lambda$ and $(Re) = Ud/\nu$ gives $\bar{h} \propto U^{0\cdot6}d^{-0\cdot4}$ where U is the wind-speed. Since the heat

8

flux $\phi_h = \bar{h}\Delta\theta$, then for a body area A, the total convective rate of heat loss $q_h = \phi_h A$, and since $A \propto d^2$, then $q_h \propto \bar{h}d^2\Delta\theta$ i.e. $q_h \propto U^{0\cdot6}d^{1\cdot6}\Delta\theta$.

At equilibrium, assuming negligible heat losses other than by convection, q_h should equal the rate of heat input to the body which was adjusted to be proportional to d^3. Hence, $U^{0\cdot6}d^{1\cdot6}\Delta\theta \propto d^3$ or $\Delta\theta \propto U^{-0\cdot6}d^{1\cdot4}$. This agrees closely with the experimentally established relationship mentioned above and confirms that with forced convection, insect bodies can be treated as spheres or cylinders.

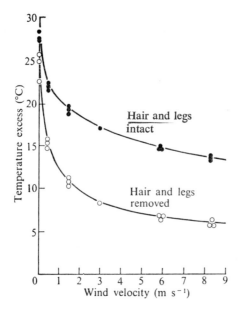

FIG. 7.9. Effect of windspeed on the temperature excess of artificially-heated normal and denuded bodies of the moth *Sphinx ligustri*. From Church (1960).

Church found that the internal excess temperature of normal insect bodies was higher than that of bodies denuded of hair, the difference increasing with wind-speed up to about 2 to 3 m s^{-1} and then remaining more or less constant at about 6 °C for the bumble bee and about 8 °C for the Privet Hawk moth (Fig. 7.9). The effect of body hair in reducing convective heat losses at relatively high wind-speeds is clearly demonstrated.

A considerable number of studies have been made on heat convection from mammalian, especially human, bodies. Because of the difficulties in maintaining a uniform air stream, few of the data can be transformed into the conventional $(\overline{Nu}) \times (Re)$ relationships for forced convection. Some data for sheep, after correcting for radiative heat exchange, are illustrated in Fig. 7.10, which indicates a relationship reasonably close to that for a cylinder of diameter, 0·3 m. For comparison, Fig. 7.10 also includes data for a nude man standing erect in a horizontal air stream; with a diameter of 0·33 m, the relationship suggested is $\phi_h = 8·5\sqrt{U}$ (W m^{-2}), i.e. about 20 per cent greater than that for the sheep.

Rapp (1970) has reviewed several studies on heat loss from human subjects in various positions. Provided that the skin is at least 10 °C warmer than the ambient air, free convection predominates at wind speeds up to about 0·5 m s^{-1} and except when mixed convection occurs, generally good agreement has been found between measured and predicted Nusselt numbers.

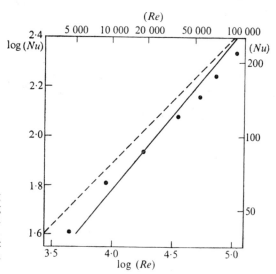

FIG. 7.10. Relation between log Nusselt number (Nu) and log Reynolds' number (Re) for forced convection from a sheep (filled circles) and human subject (pecked line). From Monteith (1973).

Body size and heat balance

According to Bergmann's rule, homeotherms (i.e. warm-blooded animals) from colder climates tend to be larger in size than their relatives from warmer regions. Though by no means a universal rule, it applies widely to birds and mammals, and is usually interpreted in terms of the heat balance of the animal. Heat loss by convection increases with increasing surface area of the body, whereas metabolic heat production increases approximately with the volume. Since the surface area/volume ratio decreases with increasing body size, a large body would tend to favour heat production whilst a small body would favour heat loss. Allen's rule, which is concerned with the marked tendency towards a reduction in the size of the body extremities in colder climates, e.g. in ear length, is clearly related to the same problem of heat balance. The above of course take no account of the insulation provided by fur on the bodies of animals in cold climates.

Convection in pipes

The exchange of heat between a circulating body fluid (e.g. blood, haemo-lymph) and the surrounding tissues is basically a problem of convective exchange between a fluid flowing in a pipe and the wall of the pipe.

With laminar flow and a fully developed parabolic velocity profile across the pipe (i.e. assuming no complicated entry effects), theory predicts that for a length of pipe l, with a uniform wall temperature, the average Nusselt number $(\overline{Nu})$ can be expressed as:

$$(\overline{Nu}) = f\left(\frac{\omega c_p}{\lambda l}\right), \tag{7.20}$$

where f signifies 'a function of', ω is the mass rate of flow along the tube (kg s^{-1}), c_p the specific heat of the fluid and λ its thermal conductivity, both these latter properties being measured at the arithmetic mean temperature of the wall and bulk fluid. The bulk temperature is defined as the temperature of the fluid when thoroughly mixed and takes account of any variation in temperature within the fluid.

If the average velocity of flow is u_m (m s^{-1}), the average fluid density ρ and the pipe diameter d, then $\omega = \pi d^2 u_m \rho/4$. Substituting this in eqn (7.20) gives:

$$(\overline{Nu}) = f\left(\frac{\pi u_m d^2 c_p \rho}{4 \lambda l}\right). \tag{7.21}$$

Multiplying top and bottom of eqn (7.21) by ν, and putting $(Re) = u_m d/\nu$ and $(Pr) = c_p \rho \nu/\lambda$ allows this equation to be expressed in the form:

$$(\overline{Nu}) = f\left(\frac{\pi}{4}\right)(Re)(Pr)\left(\frac{d}{l}\right). \tag{7.22}$$

In this form we see that the expression corresponds to the general relationship predicted by dimensional analysis for forced convection,

$$(\overline{Nu}) = f(Re)(Pr).$$

The nature of the function f can only be evaluated by experiment, by plotting the data say, in the form of eqn (7.20) i.e. $(\overline{Nu})$ $(= \bar{h}\, d/\lambda)$ against $\omega c_p/\lambda l$ on a log–log scale, as illustrated in Fig. 7.11a. For values of $\omega c_p/\lambda l > 10$, a close fit to the curve is given by the equation:

$$(\overline{Nu}) = 1.62\left(Re.Pr.\frac{d}{l}\right)^{\frac{1}{3}}. \tag{7.23}$$

A modified form of eqn (7.23) due to Seider and Tate is:

$$(\overline{Nu}) = 1\cdot86\left(Re.Pr.\frac{d}{l}\right)^{\frac{1}{3}}\left(\frac{\eta}{\eta_s}\right)^{0\cdot14} \tag{7.24}$$

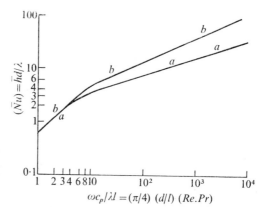

FIG. 7.11. Pipe convection. Average Nusselt numbers ($\overline{Nu}$) for developed laminar flow in tubes with a uniform wall temperature, (a) with parabolic flow profile, (b) with a rod-like or plug flow profile. Redrawn from data of Bennett and Myers (1962).

where η/η_s represents the ratio of the fluid viscosity at the bulk fluid temperature (η) to that at the wall temperature (η_s); this is an empirical correction for any distortion of the velocity profile due to variations in fluid temperature and therefore, viscosity.

Charm, Paltiel, and Kurland (1968) have confirmed the general applicability of eqn (7.23) for blood flow in glass capillary tubes by first measuring the relevant thermal properties of whole blood and plasma, and then determining heat transfer coefficients when these fluids were passed at different steady rates through tubes of known length surrounded by a constant temperature water jacket. For whole blood, the thermal conductivity λ ranged from 0·49 to 0·57 W m^{-1} K^{-1} and the specific heat c_p from 3·6 to 4·2 $\times$ 10^3 J kg K^{-1}; for plasma, the corresponding values were $\lambda = 0\cdot57$–$0\cdot61$ and $c_p = 4\cdot0$ to $4\cdot2 \times 10^3$. When $\bar{h}d/\lambda$ was plotted against $\omega c_p/\lambda l$, a curve was obtained very similar to that in Fig. 7.11, (curve a) without any need to apply the viscosity correction as in eqn (7.24).

Further studies on pulsed blood flow in the same capillaries with pulse frequencies ranging from 30 to 120 cycles s^{-1} (Hz), gave results similar to those with steady flow.

There is, however, convincing evidence that blood flowing in small capillaries cannot be treated as a uniform fluid. As will be discussed later in Chapter 10 it appears that in the terminal capillary vessels, the red cells move axially with their faces perpendicular to the line of flow. This is known as '*plug flow*' and is characterized by a more or less uniform velocity profile across the vessel; also, according to Prothero and Burton (1961; 1962), by an enhanced rate of heat exchange because of eddy motion in the plasma between the cells (see Fig. 10.5).

It is of interest that for an entirely different reason, namely, large variations in viscosity due to large radial temperature variations within the fluid, such as would occur with large temperature differences between wall and fluid, a more or less uniform velocity profile may also be developed with ordinary fluids instead of the parabolic profile assumed above. This type of flow has also been called plug flow (or rodlike flow) and the corresponding relationship between $(\overline{Nu})$ and $\omega c_p/\lambda l$ is illustrated in Fig. 7.11 (curve b), also for a uniform wall temperature. We see that when $\omega c_p/\lambda l < 10$, the curve is identical with that for parabolic flow; such a case would arise with very long pipes and slow rates of flow, when the bulk fluid temperature approximates to that of the wall. Above $\omega c_p/\lambda l = 10$ however, plug flow becomes increasingly more effective for heat transfer.

Another feature of 'pipe' flow in organisms is the relatively low thermal conductivity of the vessel walls. As in the case of convection from body surfaces therefore, (p. 97), it is unlikely that the wall temperature can be considered constant along the tube and it is more likely that we are dealing with a case of uniform heat transfer. Theory predicts that in this case, both for parabolic and plug flow, the same general relationships apply as for a uniform wall temperature, but with somewhat higher convection coefficients.

Turning to turbulent flow in pipes, it may be thought that because of the relatively low Reynolds' number of the conducting systems in organisms, (i.e. below about 2000) this would rarely occur. However, as already discussed (p. 19) there are cases where turbulence does occur in the blood system and as mentioned above, there is evidence of eddy diffusion even in the smallest capillaries, (see also p. 173).

The solutions available for turbulent heat-transfer in pipes are all empirical in origin and have been based either on experimental correlations between (Nu) and $(Re)(Pr)$, or, as will be explained later, on relationships between heat transfer and skin friction. Besides (Re) and (Pr), other factors influencing (Nu) or h are the length of the pipe and the variation in fluid properties with temperature. Some expressions attempt to account for all these factors, and assuming no entry effects, the following is applicable to both gases and biological liquids over the range $10^3 < Re < 9 \times 10^4$:

$$(\overline{Nu}) = \frac{\overline{h}d}{\lambda} = 0 \cdot 032 (Re)^{0 \cdot 8} (Pr)^n \left(\frac{l}{d}\right)^{-0 \cdot 054}, \qquad (7.25)$$

where $n = 0 \cdot 37$ when the fluid is being heated and $0 \cdot 30$ when being cooled.

Another rather better known equation, when $(l/d) > 10$, is that of Dittus and Boelter:

$$(\overline{Nu}) = 0 \cdot 023 (Re)^{0 \cdot 8} (Pr)^n, \qquad (7.26)$$

where again, $n = 0 \cdot 4$ for heated and $0 \cdot 3$ for cooled fluids.

In both the above equations, the fluid properties are those corresponding to the arithmetic average temperature of the wall and bulk fluid.

Other equations will be found in the literature, but it should be remembered that all are empirical, that the constants vary continuously over the range of (Nu), and that no single formula is best for all systems.

Counter-current heat exchange in limbs

When a mammal is exposed to cold, its limbs often become very much colder than its body. Although conduction of heat from the body is limited by the low thermal conductivity of the tissues, the limbs still require a blood supply and much valuable body heat can be lost in this way. In order to reduce heat losses due to blood flow, the outgoing arterial blood exchanges heat with the returning venous blood in a process known as *counter-current heat exchange*. To facilitate this exchange, the arteries and veins run alongside each other in the limbs, often for considerable distances. In certain species, e.g. wading and swimming birds, the system is very highly developed. Scholander (1955) has shown that the Arctic gull lost only a little more heat when its legs were placed in ice-cold water despite its web being well supplied with blood (see also, Scholander 1957).

The Reynolds' analogy

Imagine that a fluid, moving over a surface with a free stream velocity U, produces a turbulent boundary layer and that a parcel or eddy of fluid, of mass m moves from the free stream to the surface where it is brought to rest (Fig. 7.12). The momentum transferred to the surface will then be mU (i.e. momentum in the direction parallel to the surface). As we have seen earlier, the drag force, or skin friction τ_0, between fluid and surface is given by the change in momentum parallel to the surface, therefore, by mU. If the temperature of the free stream is $\Delta\theta$ °C higher than that of the surface, and the fluid particle remains at the surface long enough to attain its temperature, the amount of heat transferred will be $mc_p\,\Delta\theta$, where c_p is the specific

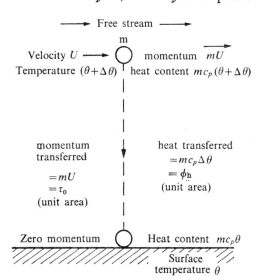

Fig. 7.12. The Reynolds' analogy. Momentum and heat transfer by eddies moving from the free stream to a surface.

heat of the fluid. The ratio of heat to momentum transfer will then be $c_p \, \Delta\theta/U$ and this will apply to any area of the surface over which both properties are transferred. If ϕ_h is the *local* rate of heat transfer per unit area of surface and τ_0 the corresponding skin friction, then

$$\frac{\phi_h}{\tau_0} = \frac{c_p \, \Delta\theta}{U}. \tag{7.27}$$

This is known as the Reynolds' analogy and rests on his original argument that in the turbulent zone, where fluid properties are transported by eddies, molecular diffusion will be negligible and there will be a proportionality between heat and momentum transport.

As described earlier (p. 41), skin friction can be expressed in terms of a local drag coefficient, $C_t = \tau_0/\frac{1}{2}\rho U^2$. Since $\phi_h = h \, \Delta\theta$, where h is the local convection coefficient, eqn (7.27) can be rewritten:

$$h = \tfrac{1}{2}C_t\rho c_p U. \tag{7.28}$$

For a plate, at a distance l from the leading edge, $(Re) = Ul/\nu$; putting $h = (Nu)\lambda/l$ and $(Pr) = c_p\rho\nu/\lambda$ transforms eqn (7.28) into:

$$(Nu) = \tfrac{1}{2}C_t(Re_l)(Pr). \tag{7.29}$$

We see therefore, that the Reynolds' analogy corresponds to the general relationship applying to forced convection, $(Nu) = f(Re)(Pr)$.

The same relationship applies to turbulent flow in pipes if U is taken as the mean velocity of flow along the pipe and h and (Re) are based on the pipe diameter d instead of l. For a pipe with turbulent flow C_t can be expressed empirically as $0.046(Re)^{-0.2}$; substituting this in eqn (7.29) gives:

$$(Nu) = 0.023(Re)^{-0.2}(Re)(Pr) = 0.023(Re)^{0.8}(Pr). \tag{7.30}$$

This would be identical with eqn (7.26) for local heat transfer in pipes if $(Pr) = 1$, i.e. if the kinematic viscosity and thermal diffusivity of the fluid were the same. The Reynolds' anology is in fact based on this assumption, so that eqn (7.30) should really be written as $(Nu) = 0.023(Re)^{0.8}$.

With the same limitation, namely $(Pr) = 1$, eqns (7.27–7.29) can also be applied to the whole surface area of a body over which momentum and heat transfer occur, but instead of the local convection and drag coefficients, we must use the average coefficients $\bar{h}$ and C_F. For a flat plate of unit width and length l_T, completely occupied by a turbulent boundary layer, eqn (4.12) gives $C_F = 0.073(Re)^{-0.2}$. Substituting this in eqn (7.29) with (Nu) now written as $\overline{(Nu)}$ and (Re) corresponding to the length l_T, gives:

$$\overline{(Nu)} = 0.036(Re)^{-0.2}(Re)(Pr) = 0.036(Re)^{0.8}(Pr). \tag{7.31}$$

Again this would correspond to the expression for turbulent heat transfer from a plate in Table 7.4 if $(Pr) = 1$.

Although the Reynolds' analogy provides a simple relationship between convective heat transfer and skin friction, it is really only applicable to fluids for which $(Pr) = 1$, though it can be used as an approximation for air $((Pr) = 0.71)$. There is however, another serious drawback to the Reynolds' analogy in that it assumes that the turbulent boundary layer extends right to the surface and therefore ignores the laminar sub-layer which lies next to the surface below the turbulent zone (see Fig. 1.7) and p. 12). Since momentum and heat can only cross this layer by molecular diffusion, this limits the rate of transfer of these properties between the surface and bulk fluid.

Several attempts have been made to modify the Reynolds' analogy by taking into account molecular diffusion across the laminar sub-layer and making it applicable to liquids such as water with (Pr) values much greater than 1. As these formulae are of interest to the engineer rather than to the biologist we need not consider them any further. The main purpose of this account is to introduce the principles underlying the analogy between momentum and heat transfer since, as we shall see later, the same principles can be extended to mass transfer and provide an important basis for analysing the transport of these three properties in the air above the ground.

8. Diffusion and convective transport of gases

So far we have only dealt with the movement of a gas (i.e. air) in terms of its bulk flow under a pressure gradient. We now turn to the movement of the constituent gases, O_2, CO_2, and water vapour between organisms and their immediate environment and that of O_2 and CO_2 in dissolved form within the tissues. The transport of O_2 and CO_2 is of course of basic importance in respiration and photosynthesis, whilst that of water vapour, between a body surface and the surrounding air, plays a key role in the water balance of the organism, and frequently also in its temperature regulation. Many of the principles discussed in this chapter apply also to the transport of other solutes in solution, but because of several complicating phenomena, this particular problem will not be discussed here. Some mention of the problem is made in Chapter 11 in relation to the transport of liquid water through tissues.

Molecular diffusion

The basic process with which we are initially concerned is that of molecular diffusion, an account of which has already been given in connection with heat conduction. In the case of the diffusion of gases, the same random movement of molecules results in a net transfer of the gas from regions of higher to lower concentration. Instead of a temperature gradient, the driving force is a concentration gradient, although (see Appendix 15), this may be expressed in many other ways, e.g. moles, pressures etc.

If we express the concentration of a substance in terms of c, its mass per unit volume (kg m^{-3}), the rate at which a mass m (kg) of that substance diffuses across an area A (m^2) of medium, perpendicular to the direction of movement, will be directly proportional to the concentration gradient in the same direction. This is Fick's first law of diffusion and is best expressed in the derivative form:

$$\frac{dm}{dt} = -DA\frac{dc}{dx} \quad \text{or} \quad \phi_c = -D\frac{dc}{dx}, \tag{8.1}$$

where the flux ϕ_c = mass of substance diffusing across unit area in unit time.

The coefficient D is known as the *diffusion coefficient*, or *diffusivity*, and has units of m^2 s^{-1}, the same as the kinematic viscosity ν and the heat diffusivity

α. Strictly speaking, a partial differential should have been used since the equation can be generalized to cover diffusion in all three dimensions, but for the sake of simplicity we shall only deal with diffusion along the x-coordinate. The negative sign simply implies that diffusion occurs in the direction of a decreasing concentration.

Diffusion in gaseous phase

The diffusivity varies with the nature of the gas, the diffusing medium, the temperature and also with the pressure.

In the simplest case, that of a gas diffusing within itself (self diffusion), kinetic theory predicts that D is proportional to the product of the mean velocity of the molecules $\bar{u}$ and their mean free path λ. It can be deduced that:

$$\bar{u} \propto \sqrt{\left(\frac{pV}{mN}\right)} \qquad \text{and} \qquad \lambda \propto \frac{V}{Nd^2} \qquad (8.2)$$

where p = gas pressure, N = number of molecules, each of mass m in volume V and d = molecular diameter.

Since mN = total mass of gas, mN/V = gas concentration or density (ρ kg m^{-3}), so that $\bar{u}$ and therefore D, will decrease as the square root the density (hence Graham's law of diffusion).

According to the universal gas law, $pV = NRT$ where R = gas constant and T = absolute temperature. Substituting into eqn (8.2) gives:

$$\bar{u} \propto \sqrt{\left(\frac{RT}{m}\right)} \qquad \text{and} \qquad \lambda \propto \frac{RT}{pd^2} \qquad (8.3)$$

At constant T therefore, $D(\propto \bar{u}\lambda)$ will be inversely related to the pressure p and to the square root of the molecular weight (m); it should also increase as $T^{3/2}$, but experiments suggest a temperature coefficient varying between 1·75 and 2.

The diffusivity of a gas is closely related to its viscosity ($D \approx \eta/\rho$) and to its thermal diffusivity ($D \approx \alpha/c_V\rho$ where c_V = specific heat at constant volume). Some values for self diffusion coefficients are given in Table 8.1. Although, even at ordinary temperatures, molecular velocities are extremely high, of the order of hundreds of m s^{-1}, the diffusivities are not so because of frequent collisions between molecules, as reflected in the relatively small values for λ (see p. 28).

In biological systems however, we are invariably concerned with mutual diffusion, i.e. the diffusion of one gas into another gas or mixture of gases such as air. Generally speaking each gas still diffuses according to its concentration gradient, but because of collisions between molecules of the different species, the mutual diffusion coefficient is influenced by the other gases present.

TABLE 8.1
Diffusivity of gases at atmospheric pressure (1.013 bar)

Temperature °C		$10^{-5}\text{m}^2\,\text{s}^{-1}$	Temperature °C		$10^{-9}\text{m}^2\,\text{s}^{-1}$
0	O_2 in air	1·78	18	O_2 in water	1·98
20	O_2 in air	2·01	25	O_2 in water	2·41
0	CO_2 in air	1·38	20	CO_2 in water	1·77
20	CO_2 in air	1·59	25	CO_2 in water	2·00
0	H_2O in air	2·20	34	CO_2 in water	1·98
20	H_2O in air	2·49	22	N_2 in water	2·02
0	CO_2 in CO_2	0·51	34	N_2 in water	2·56
0	O_2 in O_2	1·89	25	H_2O in H_2O	2·44

Data from International Critical Tables; Reid and Sherwood (1966).

For an isothermal binary system comprising two gases, 1 and 2, kinetic theory predicts that the mutual diffusion coefficient of gas 1 into gas 2 (D_{12}) is:

$$D_{12} \approx \frac{n_1 \bar{u}_2 \lambda_2 + n_2 \bar{u}_1 \lambda_1}{3(n_1 + n_2)} \tag{8.4}$$

where n_1, n_2 refer to the number of molecules of each gas per unit volume (the molecular density).

If the pressure within such a binary system is to remain uniform, the total molecular density ($n_1 + n_2$) must also be uniform; this means that the molecules of one diffusing gas must be replaced by an equal number of molecules of the other gas. In other words, the mutual diffusion coefficient for gas 2 into gas 1 (D_{21}) must be the same as D_{12}. Differences between the mutual diffusion coefficients for H_2O, CO_2, and O_2 in air at the same temperature (Table 8.1) are of course, due to differences in their molecular weights and diameters affecting $\bar{u}$ and λ.

The values given in Table 8.1 refer to the given temperatures and to a standard (total) pressure of 1·013 bar (1 atm). To convert these values to other temperatures and/or pressures, the following equation is used:

$$D_{T,p} = D_0 \left(\frac{T}{273}\right)^m \left(\frac{1\cdot013}{p}\right) \tag{8.5}$$

where D_0 refers to the tabulated value at STP (0°C and 1·013 bar) and $D_{T,p}$ to the diffusivity at an absolute temperature T and pressure p bar. For CO_2, $m = 2$, whilst for O_2 and H_2O, $m = 1\cdot75$.

Thus at atmospheric pressure, D_{0a} for O_2 diffusing in air at 0 °C = $1\cdot78 \times 10^{-5}$ m^2 s^{-1}; at 20 °C (293 K) and atmospheric pressure, $D_{0a} = 1\cdot78 \times 10^{-5} \times (293/273)^{1\cdot75} = 2\cdot01 \times 10^{-5}$ m^2 s^{-1}.

Although eqn (8.4) suggests that the mutual diffusion coefficient is very dependent on the relative proportions of the two gases, this is largely due to certain simplifying assumptions made in deriving this equation. More rigorous calculations confirm experimental evidence that the effect of gas composition is much smaller; in fact, in most biological systems, this factor can be ignored with little error.

An interesting situation may develop when two gases diffuse simultaneously in air in opposite directions, as in the case of water vapour and CO_2 diffusion through leaf stomata during transpiration and photosynthesis. Here we are dealing essentially with a ternary system involving mutual collisions between H_2O, CO_2, and air molecules. When transpiration is high, water molecules diffusing out of the leaf and colliding with CO_2 molecules may induce a net movement of the latter in the same direction, independent of the CO_2 gradient; around the compensation point therefore, when the CO_2 gradient is small or absent, the concept of a CO_2 diffusion resistance based on the ratio (gradient)/(CO_2 flux), (see p. 118) may become rather questionable (Jarman 1974).

Steady state diffusion

Eqn (8.1) applies only to the steady state, i.e. to a system in which there is no change in concentration with time, so that both the gradient dc/dx and the rate of diffusion dm/dt are constant. If at the beginning and end of a diffusion pathway of length l, the concentrations are respectively c_0 and c, the gradient will be $c_0-c/l = \Delta c/l$; assuming D is constant, eqn (8.1) can then be integrated to give:

$$q_c = \frac{m}{\Delta t} = DA\left(\frac{\Delta c}{l}\right). \tag{8.6}$$

In other words, m is the mass of substance that has diffused over an interval of time Δt across a length of medium l with a cross-sectional area A when the difference in concentration is Δc.

As an example we might take the case of the evaporation of water contained at the bottom of a tall cylinder of height l and known diameter. After a certain time a steady state will be reached when water vapour will diffuse at a constant rate through the depth of still air into the surrounding air (assumed to be at constant humidity and temperature). The concentration of the water vapour in the air just above the water surface will be that of saturated air at the prevailing water temperature $= c_s$ (kg m^{-3}); if the water vapour concentration in the air above the cylinder is c (kg m^{-3}), then $\Delta c = c_s-c$. Since D for water vapour in air is known (Table 8.1), these data can be inserted into eqn (8.6) to determine the rate of evaporation.

If we were to measure the concentration of water vapour in the air in the cylinder we would find that it decreased linearly with height, according to the relationship:

$$c = c_s-kx, \tag{8.7}$$

where x is the distance from the water surface. The constant k is the slope of the gradient $(c_s - c)/l$.

Diffusion through pores and multiperforate septa

Most biological systems in which steady state diffusion can be assumed are rather more complex than the example just discussed. In transpiration for example, water vapour diffuses from the moist surfaces of the mesophyll cells within the leaf across an internal air space (the sub-stomatal cavity), through the stomatal pores in the epidermis into the surrounding air (Fig. 8.1). The exchange of CO_2 and O_2 in photosynthesis and respiration also occurs by diffusion along the same pathway, the most important feature of which is the multiperforate nature of the epidermis.

The classic work in this field was carried out by Brown and Escombe (1900). Following measurements of the rate of CO_2 diffusion from the air into NaOH contained at the bottom of a cylinder, they sealed the cylinder with a thin septum pierced with a small circular aperture and found that the rate of diffusion was much greater than would be expected from the area of the aperture. Experiments with different sized apertures indicated that the rate of diffusion was proportional to the diameter of the aperture rather than to its area. The same behaviour was found to apply to the diffusion through small apertures of water vapour, and also of a salt in 5% gelatin.

The explanation for these findings had been anticipated theoretically by Stefan in 1881 in the case of evaporation from small circular liquid surfaces. By analogy with the electric field over a charged area, he argued that the vapour travelled along hyperbolic pathways with their foci around the perimeter of the area. Thus, instead of the planes of equal vapour concentration (i.e. density) lying parallel to the evaporating surface they formed a series of more or less concentric shells over the surface. Brown and Escombe assumed that the same shells would form above and below a small isolated circular aperture (Fig. 8.2). Now if this figure is drawn to half scale, corresponding to an aperture of half the diameter, a shell corresponding to a particular vapour concentration in the larger figure say, c_2 would appear twice as near to the surface in the smaller figure. In other words, by halving the diameter of the aperture the gradient of concentration and therefore the rate of diffusion would be doubled; hence the proposed proportionality between the rate of diffusion and the diameter of the aperture.

Fig. 8.3 illustrates the situation envisaged for a cylindrical tube corresponding to a circular stomatal pore, provided that the air above is reasonably still and that there is space enough below to accommodate the lower shell (as would probably be the case in the sub-stomatal cavity). Using a somewhat different terminology, they argued that if the vapour concentration in the bulk air well above the pore was c and that at the upper surface of the pore was c_1, the mass rate of flow through the top shell (q_c) would be given by:

$$q_c = 2Dd(c-c_1), \qquad (8.8)$$

where D = the diffusivity of the vapour in air and d = the diameter of the pore.

If the length of the pore is l and its cross-sectional area is A, and if the vapour concentration at the bottom of the pore is c_2, then according to eqn (8.6) the mass rate of flow through the tube will be:

$$q_c = DA\left(\frac{c_1-c_2}{l}\right). \qquad (8.9)$$

Since the flow rate through the shell and tube must be the same, we can

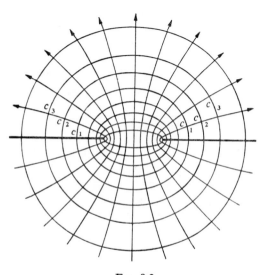

FIG. 8.2

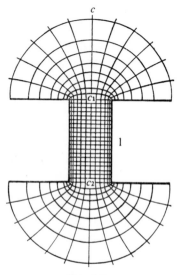

FIG. 8.3

FIG. 8.2. Hyperbolic pathways of vapour diffusion through a small circular aperture giving rise to 'shells' of equal vapour concentration, c_1, c_2 etc. From Brown and Escombe (1900).

FIG. 8.3. Formation of vapour shells above and below a narrow cylindrical pore corresponding to a stomatal pore. From Brown and Escombe (1900).

FIG. 8.4. Pathways of vapour diffusion through a multiperforate septum with planes of equal vapour concentration above and below the septum. From Brown and Escombe (1900).

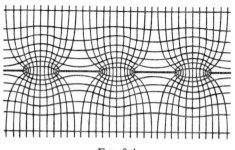

FIG. 8.4

eliminate c_1 from eqns (8.8 and 8.9) to obtain:

$$q_c = \frac{DA(c-c_2)}{l+(\pi d/8)}. \tag{8.10}$$

It follows that the presence of a vapour shell above the top of a pore is equivalent to increasing the length of the pore by an amount $\pi d/8$. If a shell is formed over the bottom of the pore, as in Fig. 8.3, a further length of $\pi d/8$ would have to be added, making a total increase of $\pi d/4$. Penman and Schofield (1951) point out an acoustic analogy known as the 'end correction' for calculating the true pitch of an organ pipe.

Brown and Escombe made further experiments on diffusion through membranes containing various numbers of small circular apertures (from 2·8 to 100 per cm²) and again found rates much higher than would be expected from the pore area; e.g. for a septum with 11·3 per cent pore area, diffusion occurred at almost the same rate as through a completely open cross-section, whilst with only 0·31 per cent pore area, the rate was still about 20 per cent. They attributed this to the steepening of the concentration gradient by vapour shells as in the case of single apertures. The situation they envisaged for a porous septum is illustrated in Fig. 8.4; hyperbolic streams of vapour leaving the pore are assumed to diverge, but when they meet streams from neighbouring pores they bend and then run parallel to each other. When the pores were sufficiently wide apart from each other for each to exercise its influence without interference (a minimum of about 10 diameters was suggested) they argued that the same conditions would apply as for a single pore, namely that the actual pore length would be increased by $\pi d/8$ at each end. Hence for a leaf epidermis with N pores per unit area, the vapour flux ϕ_c (mass flow rate per unit area of leaf surface) for a concentration difference Δc between the ends of the pore would be:

$$\phi_c = \frac{NDA\,\Delta c}{l+(\pi d/4)} = \frac{ND\pi d^2\,\Delta c}{4l+\pi d} \qquad \text{(since } A = \tfrac{1}{4}\pi d^2\text{).} \tag{8.11}$$

This follows from eqn (8.10) assuming that shells are formed at both ends of the pore.

Despite the fact that calculations of CO_2 and water vapour diffusion rates based on eqn (8.11) for given stomatal sizes and frequencies proved to be much greater than those observed in plants, Brown and Escombe's findings were to dominate plant physiologists' concepts of stomatal behaviour for a number of decades, in particular, their emphasis on pore diameter as a determining factor. Later observations by Sayre (1926) which appeared to confirm the dependence of evaporation from small open water surfaces on their diameter (or perimeter), led to the concept of the 'perimeter law' governing stomatal control of transpiration. It was argued that since the perimeter changes much less than the area when the stomatal pore closes, diffusion through slightly-closed stomata would almost be as great as when they were fully open; in other words, stomatal control would be most effective when they were nearly closed. This conflicts with the general experience that stomata regulate transpiration over the whole range of openings.

It is now generally agreed that the main error of these earlier concepts of stomatal control was due to insufficient recognition of the need to include the resistance of the air to vapour diffusion, especially as it is affected by wind. Since the movement of a vapour across the laminar boundary layer formed over leaf surfaces is only possible by molecular diffusion (much slower than in the turbulent zone above), it follows that the greater the wind-speed and the thinner the boundary layer, the higher will the rate of diffusion to and from a leaf, i.e. the lower the resistance to vapour flow.

Diffusion resistances

At this stage it is necessary to define more precisely the concept of diffusion resistance. As in the case of heat flow (p. 82) we can adopt the analogy with Ohm's law and regard the vapour flux (ϕ_c) across a diffusing medium as being 'driven' by a concentration difference (Δc) through a diffusion resistance (r) just as an electric current (I) is 'driven' by a voltage difference (V) through an electrical resistance (R).

Rewriting eqn (8.6) in the form of Ohm's law ($I = V/R$):

$$\phi_c = \frac{m}{A\,\Delta t} = \frac{\Delta c}{l/D} = \frac{\Delta c}{r}, \tag{8.12}$$

where as before, l = length of the diffusion pathway, A its cross-sectional area (n.b. the *leaf surface* area) and D = diffusivity of the vapour in air; the quantity l/D then corresponds to the diffusion resistance r with units of s m^{-1} (n.b. most texts quote r in terms of s cm^{-1}).

If as illustrated in Fig. (8.5b), we divide the transpiration pathway into two parts: (1) from the surface of the mesophyll cells to the outside surface of the leaf; and (2) from the surface of the leaf into the bulk air outside the boundary layer, we can estimate their respective diffusion resistances, r_l and r_a by comparing, under the same environmental conditions, the rate of transpiration from a leaf well supplied with water, with the rate of evaporation from a replica made of filter paper saturated with water (Fig. 8.5a).

If the concentration of water vapour in the bulk air $= c$ (kg m^{-3}) and that at the surface of the filter paper $= c_s$ (corresponding to the saturation concentration at the prevailing temperature), the evaporation flux ϕ_E (kg m^{-2} s^{-1})

9

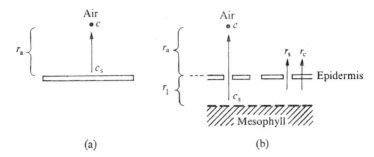

(a)   (b)

FIG. 8.5. Schematic diagram of water vapour diffusion (a) from a moist filter paper through an air resistance r_a and (b) from leaf mesophyll through a leaf resistance r_1 in series with r_a. Note that the stomatal resistance r_s is in parallel with a cuticular resistance r_c.

is given by eqn (8.12):

$$\phi_E = \frac{c_s - c}{r_a}, \qquad \text{whence } r_a = \frac{c_s - c}{\phi_E} \qquad (8.13)$$

In the case of the leaf, water vapour has to diffuse through two resistances in series, the leaf resistance r_1 plus the air resistance r_a. Assuming that the air at the surface of the mesophyll cells is saturated and that the temperature of the leaf is the same as that of the filter paper, the concentration of water vapour will be c_s, the same as for the filter paper. The transpiration flux ϕ_T (kg m^{-2} s^{-1}) is then given by:

$$\phi_T = \frac{c_s - c}{r_1 + r_a}, \qquad \text{whence } r_a + r_1 = \frac{c_s - c}{\phi_T} \qquad (8.14)$$

Assuming that r_a is the same for the leaf and filter paper replica, r_1 can be obtained by simple subtraction.

The assumptions made in the above experiment may not be entirely valid, but we need not worry about this at present since the main purpose of the experiment was to illustrate how the concept of diffusion resistance is applied. Since for a leaf, the diffusion resistance is normally dominated by the stomata, it is customary to express the data in terms of the stomatal diffusion resistance r_s rather than r_1. The latter will normally include the resistance of the pathway across the sub-stomatal chamber (usually very small) and a cuticular resistance r_c corresponding to the pathway across the epidermis (Fig. 8.5b); in most species r_c is very large, and being in parallel with the stomatal resistance, it may also be ignored.

Values for r_s will vary with species according to the size and frequency of the stomata; also, of course, on the degree of opening. Published data for fully-open stomata (Meidner and Mansfield 1968; Monteith 1973) indicate a range of about 1 to 2 s cm^{-1} for most mesophytes, to about 30 s cm^{-1} for conifers and xerophytes. The high values for conifers is due largely to the fact that the stomata are sunk below the surface of the epidermis, so increasing the length of the diffusion pathway (Fig. 8.1). In the case of amphistomatous leaves (i.e. with stomata on both surfaces) the total resistance is determined by the resistances of the two surfaces in parallel; if these are r_{s1} and r_{s2}, the total resistance r_s is given by $1/r_s = (1/r_{s1})+(1/r_{s2})$. In hypostomatous leaves the stomata normally only occur on the lower surface.

Assuming that the stomatal pores correspond to circular tubes with shells at both ends we can relate r_s to the pore dimensions and frequency by eqns (8.11 and 8.12):

$$r_s = \frac{\Delta c}{\phi_c} = \frac{4l+\pi d}{N D \pi d^2}. \tag{8.15}$$

This equation, originating from Brown and Escombe has been subsequently confirmed as being applicable to circular tubes.

For water vapour in air (Table 8.1), D at 20 °C and atmospheric pressure $\approx 2 \cdot 5 \times 10^{-5}$ m² s⁻¹. For a typical hypostomatous leaf, the pore depth $l = 30$ μm, the diameter $d = 6$ μm and the frequency $N = 3 \cdot 5 \times 10^8$ per m² leaf surface, whence $r_s = 140$ s m⁻¹ or $1 \cdot 4$ s cm⁻¹, thus of the same order as measured.

Stomatal pores however, are not circular. Most mesophytes have elliptical pores, but according to Penman and Schofield (1951) they can be treated as circular with a diameter equal to the geometric mean lengths of the long and short axes; if the semi-axes of the ellipse are a, b, then $d = ab/4$. For other shapes, including the rectangular slit like pores characteristic of the Gramineae, Parlange and Waggoner (1970) suggest:

$$r_s = \frac{l/\pi ab + \ln(4a/b)/\pi a}{ND}, \tag{8.16}$$

where a, b, and l are respectively the semi-length, semi-width and depth of the stomatal pore. These authors also state that if the interstomatal spacing is at least three times the stomatal length (as appears to be the case with most leaves) mutual interference is negligible.

As already indicated, the air resistance r_a will vary with the depth of the laminar boundary layer, which according to eqn (4.1) increases as the square root of the length of leaf l in the direction of wind movement and inversely as the square root of the wind velocity U. Instead of making allowance for the varying depth of the boundary layer over the leaf, we might take the empirical relation suggested by Monteith (1965):

$$r_a = 1 \cdot 3 \times 10^{-2} \sqrt{(l/U)} \text{s m}^{-1}. \tag{8.17}$$

Hence for a leaf with say $l = 0 \cdot 1$ m, r_a will vary from ∞ in completely still air to about 43 s m⁻¹ or $0 \cdot 43$ s cm⁻¹ at $U = 9$ m s⁻¹ (about 20 miles h⁻¹).

A comparison between r_a and r_s helps to explain why earlier experiments appeared to support the perimeter law for stomatal control of transpiration. We have seen (eqn 8.14) that transpiration is inversely related to r_1+r_a or r_s+r_a if we use only the stomatal resistance. If measurements are made in the laboratory in fairly still air, r_a will be so large that increasing r_s (i.e. by closing the stomata) will have relatively little effect on r_s+r_a until the stomata are nearly closed. In the open, with wind, r_a usually falls to a value similar to or below that of r_s, in which case even small changes in r_s will have a significant effect on r_s+r_a and hence on transpiration.

Although most of the examples described above refer to the diffusion of water vapour in transpiration, it will be appreciated that the same principles apply to the diffusion of CO_2 and O_2 through the stomata. In the case of CO_2 which has a lower diffusivity in air than water vapour (1.6×10^{-5} m^2 s^{-1} at 20 °C) it follows that both r_s and r_a will be corresponding larger (but see p. 111).

The problem of gaseous diffusion to and from plant canopies in the open will be discussed in Chapter 9.

Diffusion in liquids

Diffusion in liquids is a very much slower process than in the gaseous phase. As is evident from Table 8.1 the diffusivity of gases in water is about 10^4 times less than in air. This can be explained if we think of the movement of molecules in solution as being subject to friction drag due to the viscosity of the solvent. In fact, the problem of diffusion in liquids can be analysed on the basis of Stokes' law for small particles (p. 52); the larger the diffusing molecule and the more viscous the medium, the greater will be the drag and the lower the diffusivity (compare the values for O_2 and CO_2 in water in relation to their molecular weights).

As in the gaseous phase, the diffusivity of a dissolved gas increases with the temperature, but the situation is complicated by a reduction in solubility. Although to a certain extent these effects balance each other, there is no strict relation between the two. Since for most practical purposes, O_2, CO_2, and N_2 can be assumed to obey Henry's law, their solubility will be proportional to the pressure of the gas in the gaseous phase with which the solution is in equilibrium. In quoting solubilities therefore, it is necessary to define this pressure, or in the case of gas mixtures, the partial pressure of the gas concerned. It should be remembered that each gas constituent of a mixture dissolves independently according to its partial pressure.

The solubility of a gas may be expressed in various ways and since this can lead to confusion in applying the diffusion equations, details are given in Appendix 15, including molar concentrations and partial pressures. The values given in Table 8.2 for O_2, CO_2, and N_2 in water, and for O_2 in certain other liquids at different temperatures refer to the so-called *Bunsen coefficient* (α). This is defined as the volume of gas at STP (i.e. at $T = 273$ K and atmospheric pressure) dissolved in unit volume of liquid when the pressure of the gas in the gaseous phase is atmospheric (1.013 bar). For continuity, these volumes are quoted in m^3, but as we are dealing with a volume/volume ratio any other unit can be used. If the partial pressure of the gas in the gaseous phase is p bar, the volume of dissolved gas (still at STP) may be calculated by multiplying α by $p/1.013$ (Appendix 15). It will be noted from Table 8.2 that the solubility of O_2 is lower in salt solutions than in water. Non-electrolytes such as sucrose and glucose also reduce its solubility, probably

TABLE 8.2

Solubilities of gases in water and solutions

Temperature °C	Bunsen coefficients m^3 gas at STP per m^3 liquid in equilibrium with gas in gaseous phase at a pressure of 1 atmosphere (1.013 bar)				
	CO_2 in water	N_2 in water	O_2 in water	O_2 in sea water	O_2 in blood
0	1·713	0·0239	0·0489	0·0391	—
10	1·194	0·0196	0·0379	0·0313	—
20	0·878	0·0164	0·0309	0·0260	—
30	0·665	0·0138	0·0282	0·0219	—
35	0·592	—	0·0261	—	—
37	—	—	0·0238	—	0·019 to 0·022
40	0·530	0·0118	0·0231	—	—

Data from Dixon (1951). Handbook of Chemistry and Physics (Chemical Rubber Pub. Co.). 43rd ed.

because of solvation; according to Leonard (1939), a 10% sucrose solution reduces the O_2 solubility by about 20 per cent at 15 °C. On the other hand, provided that no chemical combination occurs, the solubility of CO_2 is usually practically independent of salt concentration. However, Leonard (1939) has shown a linear fall in CO_2 solubility with increasing sugar concentration; a 10% sucrose solution at 20 °C lowered the solubility by about 8 per cent.

Because of the different partial pressures of O_2, CO_2, and N_2 in air, and their differing solubilities, the composition of air dissolved in water is very different from that in the gaseous phase. As shown in Appendix 15, if air at atmospheric pressure and containing by volume, 79·0% N_2, 20·96% O_2, and 0·04% CO_2 is shaken up with water at 20 °C, the volume composition of the dissolved air will be 65·4% N_2, 32·8% O_2 and 1·76% CO_2. Leonard (1939) gives a useful table of O_2 and CO_2 contents of water in equilibrium with air containing 79·04% N_2 and 21% ($O_2 + CO_2$ in varying proportions).

Non-steady state diffusion

In steady state diffusion, the concentration of a diffusing substance at any point in the medium does not change with time. In many biological systems however, this is not the case. For example, when O_2 or CO_2 diffuse through a soil, they will vary in concentration as a result of the respiration of microorganisms; respiration in tissues will produce a similar effect. The situation is analogous to the non-steady diffusion of heat (p. 84) and an appropriate equation for mass (i.e. gas) diffusion can be derived along similar lines (Appendix 16):

$$\frac{dc}{dt} = D \frac{d^2c}{dx^2}, \qquad (8.18)$$

where dc/dt represents the rate of consumption or production of the diffusing substance per unit volume of medium ($kg\ m^{-3}\ s^{-1}$) and d^2c/dx^2 the rate at which the concentration gradient dc/dx changes with distance x along the direction of diffusion. Again, this equation can be arranged to deal with diffusion along other coordinates and strictly speaking, should be expressed in partial differentials.

Eqn (8.18) is known as *Fick's second law of diffusion* and like his first, assumes that D is constant; this is a fair assumption in the case of gases at relatively low pressures as in air; it does not necessarily apply to solutes in general.

Diffusion through tissues

The diffusion of O_2 and CO_2 through tissues is of fundamental importance in respiration and photosynthesis. A particularly important practical aspect in respirometry is the supply of O_2 to cells lying deep within a tissue.

Imagine a slice of tissue of thickness l suspended in a medium containing dissolved O_2 at a uniform concentration c_0 ($kg\ m^{-3}$). As the O_2 diffuses through the tissue and is consumed in respiration, its concentration will decrease. For simplicity, let us make the common assumption that the rate of respiration q (kg O_2 per m^3 tissue per sec) is uniform within the tissue and that it is unaffected by the O_2 concentration until the latter falls below a certain level; at this stage, there will be no further change in O_2 concentration with depth. If the thickness l is such that this stage is just reached at the centre of the slice (depth $l/2$), it can be deduced from eqn (8.18) (see Appendix 16) that the O_2 concentration c at any other depth (x) from the surface of the slice will be:

$$c = c_0 - \frac{qx}{2D}(l-x), \qquad (8.19)$$

where D = diffusion coefficient of O_2 in the tissue.

At the centre of the slice ($x = l/2$), the O_2 concentration c' is then given by:

$$c' = c_0 - \frac{ql^2}{8D} \quad \text{or,} \quad l = \sqrt{\left\{\frac{8D(c_0-c')}{q}\right\}}. \tag{8.20}$$

As l increases so c' decreases until for a thickness $l = l'$, it becomes zero; in which case,

$$l' = \sqrt{\left(\frac{8Dc_0}{q}\right)}; \tag{8.21}$$

l' then corresponds to the maximum thickness of tissue slice that allows some O_2 to penetrate to the deepest cells at the centre.

Because of difficulties in determining gas concentrations in tissues, it is usually not possible to evaluate the above equations directly. Instead of D, a permeability or *invasion coefficient* D' is used; this is defined as the volume rate of flow of a gas at a given temperature through unit area of tissue under a given pressure gradient. For example, in the case of O_2 diffusion through muscle tissue at 37 °C, Krogh obtained a value of $1\cdot4\times10^{-5}$ ml O_2 per minute per cm^2 tissue with a pressure gradient of 1 atmosphere per cm ($= 100$ atm per m) across the tissue. In S.I. units this is equivalent to $2\cdot3\times10^{-11}$ m^3 O_2 per m^2 tissue per second for a pressure gradient of 1 bar per m, or $2\cdot3\times10^{-11}$ m^2 s^{-1} bar^{-1}. Since the diffusion coefficient D has units of m^2 s^{-1}, then $D' = D/\Delta p$ where Δp is the pressure difference across the tissue. Only when $\Delta p = 1$ does $D' = D$.

In practice, q is usually measured in terms of a volume of O_2 consumed per unit volume of tissue in unit time (e.g. m^3 m^{-3} s^{-1} or, s^{-1}). If at the same time we use the invasion coefficient $D'(= D/\Delta p)$ instead of the diffusion coefficient D, then to preserve the dimensional balance of eqns (8.20) and (8.21), the gas concentrations c_0, c', must be expressed in terms of their equivalent pressures or partial pressures (e.g. in bars).

In eqn (8.20) the pressure equivalent of c_0-c' then corresponds to the pressure difference Δp across the depth $l/2$ of tissue. This avoids the problem of determining gas concentrations in tissues. Furthermore, if c_0 is expressed in terms of the pressure or partial pressure of O_2 in the gaseous phase, in equilibrium at the given temperature with the O_2 in the medium bathing the tissue, and c' in terms of the equivalent partial pressure of O_2 in the tissue, this largely eliminates unknown temperature effects on the diffusivity and solubility of O_2 in the tissue. In other words, D' can be assumed in practice to be more or less independent of the temperature.

In the case of eqn (8.21), since c' is zero at the centre, the pressure equivalent of c_0 corresponds to the pressure difference across the depth $l'/2$ so that we can rewrite this equation in the form:

$$l' = \sqrt{\left(\frac{8D'p_0}{q}\right)}, \tag{8.22}$$

where as before, D' = invasion coefficient ($m^2\,s^{-1}\,bar^{-1}$), p_0 = external O_2 pressure or partial pressure (bar), and q is in units of $m^3\,m^{-3}\,s^{-1}$; l is then given in m.

Taking Krogh's value of D' converted to S.I. units = $2.3 \times 10^{-11}\,m^2\,s^{-1}\,bar^{-1}$ and assuming that it applies to say, liver tissue which has a relatively high O_2 consumption of $q = 8.3 \times 10^{-4}\,m^3\,m^{-3}\,s^{-1}$, we obtain from eqn (8.22) $l' = 4.7 \times 10^{-4}\,m$ (= 0.47 mm) when in equilibrium with pure O_2 ($p_0 = 1$ bar) and $l' = 2.1 \times 10^{-4}\,m$ (= 0.21 mm) when in equilibrium with air with a partial O_2 pressure, $p_0 = 0.2$ bar.

Somewhat similar calculations have been made by Briggs and Robertson (1948) for both O_2 consumption and CO_2 production in carrot discs.

Fenn (1927) has derived an equation for a cylinder of respiring tissue which can be expressed in the form:

$$p = p_0 - \frac{q}{4D'}(r^2 - l^2), \tag{8.23}$$

where D' = invasion coefficient, r = cylinder radius, $p = O_2$ pressure within the tissue at a radial distance l from the axis and p_0 = external O_2 pressure.

He calculated that in a frog's nerve, with $r = 1$ mm and an O_2 consumption $q = 2.05 \times 10^{-5}\,cm^3$ per cm^3 tissue per second, the O_2 pressure at the axis would be just about zero in equilibrium with air. (Check by using Krogh's value for $D' = 2.3 \times 10^{-11}\,m^2\,s^{-1}\,bar^{-1}$, $q = 2.05 \times 10^{-5}\,m^3\,m^{-3}\,s^{-1}$, $r = 10^{-3}$ m and $p_0 = 0.21$ bar).

For uniformly respiring spherical cells, Gerrard (1931) derived the equation:

$$p = p_0 - \frac{q}{6D'}(r^2 - l^2), \tag{8.24}$$

where as before, r = radius and p, p_0 the respective O_2 pressures at a radial distance l from the centre and outside .

To maintain the partial pressure of O_2 at the centre ($l = 0$) just above zero, the external O_2 pressure (p_0) must be at least $qr^2/6D'$ or r not greater than $\sqrt{(6D'p_0/q)}$. In water, in equilibrium with air with p (O_2) = 0.2 bar, the radius of a spherical cell with an O_2 consumption of $1.2 \times 10^{-5}\,mm^3$ per mg fresh weight per second, must be just under 1 mm, otherwise diffusion cannot supply the necessary O_2. (Check by inserting Krogh's value for $D' = 2.3 \times 10^{-11}\,m^2\,s^{-1}\,bar^{-1}$, $p_0 = 0.2$ bar and $q = 1.2 \times 10^{-5}\,m^3\,m^{-3}\,s^{-1}$ assuming the density = 1.)

Gerrard also derives equations for cases where it cannot be assumed that respiration is uniform within a cell or independent of the O_2 concentration.

Convective transport

Just as heat may be transported by a moving fluid in the process of convection, so may solutes and gases. In fact, one of the aims of this account is to show how many of the problems of convective mass transport can be solved by a simple extension of the principles described earlier for heat convection. We shall deal first with convective transport between a body surface and the surrounding medium (air); though most of the examples refer to water vapour, the same underlying principles apply to the transport of other gases. Convection by wind above the ground will be discussed in Chapter 9.

Convection between a body surface and air

By analogy with eqn (7.16) for heat transfer, we can construct an equation for mass transfer:

$$\phi_c = h_c \Delta c, \tag{8.25}$$

where ϕ_c corresponds to the flux of material (kg m^{-2} s^{-1}), expressed as a simple product of the concentration difference Δc (kg m^{-3}) and a mass transfer coefficient h_c (units, m s^{-1}). It will be noted that the units of h_c are different from those of h.

We saw from eqn (7.19) that forced convection of heat could be expressed in terms of a general relationship of the form, $(\overline{Nu}) = c(Pr)^m(Re)^n$. To derive a comparable relationship for mass transfer we need the equivalents of (Nu) and (Pr); (Re) will of course remain the same.

(Nu) is defined as hl/λ and is dimensionless. If we retain l to characterise the surface dimension, and substitute h_c for h, then since the product $h_c l$ has dimensions $LT^{-1}L = L^2T^{-1}$, we must replace λ by D which also has the dimensions L^2T^{-1}. The new dimensionless number is known as the *Sherwood number* (Sh) and is defined as:

$$(Sh) = \frac{h_c l}{D} ; \tag{8.26}$$

(Pr) is defined as ν/α and since D is dimensionally equivalent to α, the equivalent to (Pr) for mass transfer, known as the *Schmidt* number (Sc) is given by:

$$(Sc) = \frac{\nu}{D} . \tag{8.27}$$

By analogy with eqn (7.19), the general relationship for mass transfer by forced convection can now be expressed in the form:

$$(Sh) = c(Sc)^m(Re)^n. \tag{8.28}$$

A convenient way to determine the unknown constants is to measure the rate of evaporation (corresponding to the convective transport of water vapour) from variously shaped body surfaces. Several such measurements have been made including some on leaf models (e.g. Thom 1968). For forced convection with laminar flow, the results generally agree closely enough with the equations established for heat transfer for the latter to be used with reasonable confidence for mass transfer.

The following example illustrates the procedure for calculating the rate of evaporation from a flat rectangular surface, 0·1 m long and 0·05 m wide, covered with a film of water over which air at 20 °C and with a relative humidity of 30 per cent is blown at a uniform speed of 9 m s⁻¹, parallel to the longer side.

With v for air at 20 °C $= 1·5 \times 10^{-5}$ m² s⁻¹,

$$(Re) = (9 \times 0·1)/(1·5 \times 10^{-5}) = 6 \times 10^4.$$

Flow should therefore be laminar and the equation for mass transfer, corresponding to that for heat transfer (Table 7.4) will be:

$$(\overline{Sh}) = 0·664(Sc)^{\frac{1}{3}}(Re)^{\frac{1}{2}}, \tag{8.29}$$

[note the use of the average $(\overline{Sh})$ corresponding to $(\overline{Nu})$].

From Table 8.1, D for water vapour in air at 20 °C $= 2·5 \times 10^{-5}$ m² s⁻¹ whence $(Sc)(= v/D) = 0·6$ and $(\overline{Sh}) = 137·2$.

From eqn (8.26), $\bar{h}_c = D(\overline{Sh})/l = 0·034$ m s⁻¹.

Assuming that the air next to the surface of the plate is saturated, its vapour concentration (i.e. its density) at 20 °C $= 17·3 \times 10^{-3}$ kg m⁻³ (Table 8.3), and at 30 per cent relative humidity, about $5·2 \times 10^{-3}$ kg m⁻³. Hence Δc, the difference between the surface and bulk air $= (17·3 - 5·2) \times 10^{-3} = 12·1 \times 10^{-3}$ kg m⁻³.

From eqn (8.25) putting $\phi_E = \phi_o$ to represent the flux of water vapour, we have:

$$\phi_E = \bar{h}_c \Delta c = 4·1 \times 10^{-4} \text{ kg m}^{-2} \text{ s}^{-1}.$$

Since the area of the plate $= 0·005$ m², the rate of evaporation from the *single* surface $= 2·05 \times 10^{-6}$ kg s⁻¹.

If we compare eqns (8.25) and (8.12) we see that the average mass transfer coefficient $\bar{h}_c$ corresponds to the inverse of the air diffusion resistance r_a. In the above example therefore,

$$r_a = \frac{1}{0·034} = 29·4 \text{ s m}^{-1} \quad \text{or,} \quad 0·294 \text{ s cm}^{-1}.$$

Using the same relationship, $r_a = 1/\bar{h}_c = l/(\overline{Sh})D$ we can transform eqn (8.29) into:

$$r_a = \frac{1·51}{\sqrt{(v)}} \cdot \left(\frac{l}{U}\right)^{\frac{1}{2}} \left(\frac{D}{v}\right)^{-\frac{2}{3}} \text{ s m}^{-1}. \tag{8.30}$$

From wind tunnel measurements on metal replicas of bean leaves at windspeeds greater than about 1 m s⁻¹, Thom (1968) established exactly the same relationship, but with a minor difference in the numerical constant. He showed that if D was replaced by the thermal diffusivity α, the same equation applied to the diffusion resistance (he called this a *leaf transfer resistance*) for heat. A significant feature of Thom's results was that whilst heat and mass

TABLE 8.3

Density of pure water vapour at saturation over water

Temperature °C.	10^{-3} kg m^{-3}									
	0·0	0·1	0·2	0·3	0·4	0·5	0·6	0·7	0·8	0·9
5	6·797	6·842	6·887	6·933	6·979	7·025	7·071	7·118	7·165	7·212
6	7·260	7·307	7·355	7·404	7·452	7·501	7·550	7·600	7·649	7·699
7	7·750	7·801	7·851	7·902	7·954	8·006	8·058	8·110	8·163	8·216
8	8·270	8·324	8·377	8·431	8·485	8·540	8·595	8·650	8·706	8·762
9	8·819	8·875	8·932	8·989	9·046	9·104	9·163	9·221	9·280	9·339
10	9·399	9·459	9·519	9·579	9·641	9·702	9·763	9·825	9·887	9·949
11	10·01	10·08	10·14	10·20	10·27	10·33	10·40	10·46	10·53	10·59
12	10·66	10·73	10·79	10·86	10·93	11·00	11·07	11·14	11·21	11·27
13	11·35	11·42	11·49	11·56	11·63	11·70	11·77	11·85	11·92	11·99
14	12·07	12·14	12·22	12·29	12·37	12·44	12·52	12·60	12·67	12·75
15	12·83	12·91	12·99	13·07	13·14	13·23	13·31	13·39	13·47	13·55
16	13·63	13·72	13·80	13·88	13·97	14·05	14·14	14·22	14·31	14·39
17	14·48	14·57	14·65	14·74	14·83	14·92	15·01	15·10	15·19	15·28
18	15·37	15·46	15·55	15·65	15·74	15·83	15·93	16·02	16·12	16·21
19	16·31	16·41	16·50	16·60	16·70	16·80	16·90	17·00	17·10	17·20
20	17·30	17·40	17·50	17·60	17·71	17·81	17·91	18·02	18·12	18·23
21	18·34	18·44	18·55	18·66	18·77	18·88	18·99	19·10	19·21	19·32
22	19·43	19·54	19·65	19·77	19·88	20·00	20·11	20·23	20·34	20·46
23	20·58	20·70	20·81	20·93	21·05	21·17	21·29	21·42	21·54	21·66
24	21·78	21·91	22·03	22·16	22·28	22·41	22·54	22·66	22·79	22·92
25	23·05	23·18	23·31	23·44	23·58	23·71	23·84	23·97	24·11	24·24
26	24·38	24·52	24·66	24·79	24·93	25·07	25·21	25·35	25·49	25·63
27	25·78	25·92	26·06	26·21	26·35	26·50	26·65	26·79	26·94	27·09
28	27·24	27·39	27·54	27·69	27·85	28·00	28·15	28·31	28·46	28·62
29	28·78	28·93	29·09	29·25	29·41	29·57	29·73	29·89	30·05	30·22
30	30·38	30·55	30·71	30·88	31·05	31·22	31·38	31·55	31·72	31·89
31	32·07	32·24	32·41	32·59	32·76	32·94	33·11	33·29	33·47	33·65
32	33·83	34·01	34·19	34·38	34·56	34·74	34·93	35·11	35·30	35·49
33	35·68	35·87	36·06	36·25	36·44	36·63	36·83	37·02	37·22	37·41
34	37·61	37·81	38·01	38·21	38·41	38·61	38·81	39·01	39·22	39·42

From Smithsonian Meteorological Tables.

transfer could be treated in the same way, momentum transfer could not. It is also of interest that whereas the drag on the leaf model was increased when it was held at an angle to the wind (as would be expected), mass transfer (e.g. evaporation) was hardly affected.

With very steep water-vapour gradients between a wet surface and air, sufficiently large changes in density may be established for free convection to occur; for most practical purposes however, free convection of mass can be ignored.

The Lewis relation

It will be noted (Table 8.1) that at 20 °C, D for water vapour in air $\approx 2.5 \times 10^{-5} \text{ m}^2 \text{ s}^{-1}$ and for O_2 in air, $= 2.01 \times 10^{-5} \text{ m}^2 \text{ s}^{-1}$, whilst at the same temperature, α for air is $2.13 \times 10^{-5} \text{ m}^2 \text{ s}^{-1}$. The similarity of these values suggests that D and α can be treated as equal without appreciable error; in which case, $(Sc)(= \nu/D) \approx Pr(= \nu/\alpha)$ and $(Nu) \approx (Sh)$. Since $(Nu) = hl/\lambda = hl/\alpha\rho c_p$ and $(Sh) = h_c l/D$ it follows that:

$$h_c \approx \frac{h}{\rho c_p}. \tag{8.31}$$

This is known as the *Lewis relation* and suggests that when $(Sc/Pr) \approx 1$ (this ratio is called the *Lewis number*), the mass transfer coefficient h_c can be very simply derived from the heat transfer coefficient h. The Lewis relation does not apply to CO_2 for which D in air is $1.59 \times 10^{-5} \text{ m}^2 \text{ s}^{-1}$; in such cases the following approximate relation may be used:

$$h_c \approx \frac{h}{\rho c_p} (Pr)^{\frac{2}{3}}. \tag{8.32}$$

Since the Lewis relation applies to local, besides overall transfer, it is useful in practice for identfying local variations in heat transfer from local rates of evaporation.

Simultaneous transport of heat and water vapour

To convert a liquid into its vapour at the same temperature energy is required in the form of heat, the well known *latent heat of vaporization, L*; for water, this ranges from $2.5 \times 10^6 \text{ J kg}^{-1}$ at 10 °C to 2.38×10^6 at 50 °C; at 20 °C, $L = 2.45 \times 10^{-6} \text{ J kg}^{-1}$. A given flux of water vapour therefore, will be accompanied by a corresponding flux of latent heat; the latter can be released as sensible heat when the vapour condenses back to liquid. In the worked example for evaporation from a plate, latent heat was not mentioned and it was assumed that the plate surface and bulk air were at the same temperature. In actuality, the necessary latent heat would have been abstracted from both the plate and the air.

As heat was taken from the plate, its surface temperature would have fallen below that of the air, a temperature gradient would then have been established and heat would have been transferred by forced convection from the air to the surface. Eventually therefore, a balance would have been set up in which the outflow of latent heat was balanced by the income of sensible heat and the surface temperature would become constant. If this surface equilibrium temperature was θ_0 and the temperature of the bulk air stream, assumed unchanged, was θ_a, the flux of sensible heat ϕ_h would be (eqn 7.16):

$$\phi_h = \bar{h}(\theta_a - \theta_0) = \bar{h}(\Delta\theta).$$

If the moisture concentration of the saturated air next to the surface was c_{s0} and that of the bulk air c_a, the flux of water vapour ϕ_E would be:

$$\phi_E = \bar{h}_c(c_{s0}-c_a)$$

and the corresponding flux of latent heat would be $L\phi_E$.

Since at equilibrium, $\bar{h}(\Delta\theta) = L\phi_E$, we have:

$$\bar{h}(\Delta\theta) = L\bar{h}_c(c_{s0}-c_a) \quad \text{and} \quad \Delta\theta = \frac{L\bar{h}_c}{\bar{h}}(c_{s0}-c_a). \tag{8.33}$$

The above argument can be used to explain the principle of the wet and dry bulb hygrometer. In this, air is passed over two identical thermometers one of which is sheathed in a wet cloth supplied with water. If the air is unsaturated, evaporation from the wet bulb will lower its temperature relative to that of the dry bulb; assuming that the wet bulb reads the temperature of the cooled air passing over it and that the dry bulb reads the bulk air temperature, the difference $\Delta\theta$ is given by eqn (8.33) above. For water vapour in air, the Lewis relation eqn (8.31) can be used so that eqn (8.33) can be written:

$$\Delta\theta = \frac{L}{\rho c_p}(c_{s0}-c_a). \tag{8.34}$$

where c_{s0} now refers to the saturation vapour pressure of air at the temperature of the wet bulb (θ_w).

In practice it is more convenient to express the air humidity in terms of the specific humidity q or vapour pressure, e. As explained in Appendix 15, $c = \rho_v$, the absolute humidity and $q = \rho_v/\rho_a$ where ρ_a is the density of moist air. Also,

$$q \approx 0.622\, e/p$$

where $p =$ prevailing air pressure. Hence,

$$c = q\rho_a \approx 0.622\, \rho_a e/p.$$

Substituting for c in eqn (8.34) and putting $\rho_a = \rho$ gives:

$$\Delta\theta = \frac{L}{c_p} \cdot \frac{0.622}{p}(e_{s0}-e_a).$$

If we replace e_{s0} by e_{sw}, the saturation vapour pressure of air at the temperature of the wet bulb, we get:

$$e_a = e_{sw} - \frac{pc_p}{0.622L}(\Delta\theta) = e_{sw} - \gamma(\Delta\theta). \tag{8.35}$$

The quantity $\gamma = pc_p/0.622L$ is known as the *psychrometric constant* and as we see, directly relates the vapour pressure difference $(e_{sw}-e_a)$ to the temperature difference between the dry and wet bulb $(\Delta\theta)$.

Putting $p = 1.013$ bar, $c_p = 1.007 \times 10^3$ J kg^{-1} and $L = 2.45 \times 10^{-6}$ J kg^{-1} gives $\gamma = 0.664 \times 10^{-3}$ bar K^{-1} or 0.664 mbar K^{-1}. Hence,

$$e_a = e_{sw} - 0.664(\Delta\theta) \qquad (e_a \text{ and } e_{sw} \text{ in mbar}). \tag{8.36}$$

The above equation based on the Lewis relation assumes that the diffusivities for heat (α) and water vapour (D) are the same. This is not strictly true since the psychrometric constant decreases to a constant value as the wind-speed increases. In the Assmann hygrometer, often used as a standard instrument, air is drawn over a standard size bulb at a speed of about 3 m s^{-1}; at this speed, the value $\gamma = 0.664$ mbar K^{-1} is acceptable in practice.

Convective transport within organisms

Since molecular diffusion is such a relatively slow process, the transport of solutes and gases over long distances within organisms is usually dependent on the movement of the medium itself under the action of some external force. Examples are to be found in the xylem and phloem conducting systems of plants and in the blood and respiratory systems of animals.

If the concentration of the solute is c (kg m^{-3}) and the velocity of the fluid is U (m s^{-1}), the convective component will be cU (kg m^{-2} s^{-1}). This supplements the normal diffusion rate, $-D\dfrac{dc,}{dx}$ so that the total diffusive flux ϕ_c (kg m^{-2} s^{-1}) is given by:

$$\phi_c = -D\frac{dc}{dx} + cU, \tag{8.37}$$

where as before, the negative sign indicates a decreasing concentration with increasing distance x in the direction of diffusion.

In most cases however, solute is continuously being withdrawn or added to the fluid by the surrounding tissues; we then have a system similar to that already described for non-steady state diffusion (p. 120). If solute is withdrawn from the fluid at a uniform rate, dc/dt, it can be shown that:

$$\frac{dc}{dt} = D\frac{d^2c}{dx^2} - U\frac{dc}{dx}. \tag{8.38}$$

If at time $t = 0$, at a distance x_0 along the flow pathway, the solute concentration is c_0, then the solute concentration at a distance x after time t (s) is given by:

$$c = c_0 \exp\left[\frac{U}{2D}(x - x_0) - \frac{U^2 t}{4D}\right]. \tag{8.39}$$

Differentiation of c with respect to t and x will confirm that eqn (8.39) satisfies eqn (8.38) and we see that when $x = x_0$ and $t = 0$, $c = c_0$ (since $e^0 = 1$).

When, as is usually the case, solute is transported through a tubular conducting system and there is continuous exchange through the walls, e.g. in the exchange of O_2 and CO_2 between the blood capillaries and the surrounding tissues, the problem is then similar to that of the exchange of heat in tubes (p. 102). Basically therefore, the same solutions can be used. With laminar flow for example, we can apply eqn (7.23) if $(\overline{Nu})$ is replaced by $(\overline{Sh})$ and (Pr) by (Sc). Then, assuming a constant wall concentration (i.e. a constant tissue concentration in equilibrium with the walls), we can construct curves similar to those in Fig. 7.11 for laminar or plug flow by plotting $(\overline{Sh})$ against $\omega/\rho Dl$ (the equivalent of $\omega c_p/\lambda l$).

For turbulent flow, measurements of evaporation from wet walls into air flowing along the tube have suggested the following empirical equation:

$$(\overline{Sh}) = 0 \cdot 023 (Sc)^{\frac{1}{3}} (Re)^{0 \cdot 8}. \tag{8.40}$$

9. The environment above the ground

THE aim of this chapter is to describe the interrelationships between organisms or communities living above the ground and the physical factors of the environment associated mainly with air movement. The underlying principles are essentially those of micrometeorology and microclimatology and several texts are available, some of which (e.g. Gates 1962; Geiger 1965; Lowry 1969; Monteith 1973) particularly emphasize the biological significance of of the subject.

Particular attention will be given in this chapter to the heat and water balances of organisms and communities and the ways in which these may be studied quantitatively. Most of the underlying principles have been covered in previous chapters except for radiation; although, strictly speaking, much of the following discussion on radiation lies outside the context of fluid behaviour, it plays such a key role in the heat balance that its inclusion is essential.

Radiation

All matter above absolute zero radiates electromagnetic energy with maximum emission at a wavelength (λ_{max}) inversely proportional to the absolute temperature (T) as described by Wien's law:

$$\lambda_{max} = 2897/T \ (\mu m). \tag{9.1}$$

Thus, from the knowledge that maximum emission in the solar spectrum occurs at $\lambda = 0.45 \ \mu m$ (blue–green), the temperature of the sun can be estimated as about 6000 K. The spectral distribution of solar radiation is illustrated in Fig. 9.1; almost 99 per cent of the radiant energy lies between $0.15 \ \mu m$ and $4 \ \mu m$ and is conveniently divided into three main regions—ultraviolet (u.v.) below $0.4 \ \mu m$, visible light between $0.4 \ \mu m$ (blue) and $0.7 \ \mu m$ (red), and infrared (i.r.), above $0.7 \ \mu m$.

The spectral distribution of radiation from bodies at normal biological temperatures (0–50 °C) is of the same general shape, but with maximum emission between 9.6 and $10.6 \ \mu m$ and almost all within the range 4 to 30 μm (Fig. 9.1). It is convenient to define this spectral range as long wave (LW) or thermal radiation to distinguish it from short wave (SW) radiation from the sun.

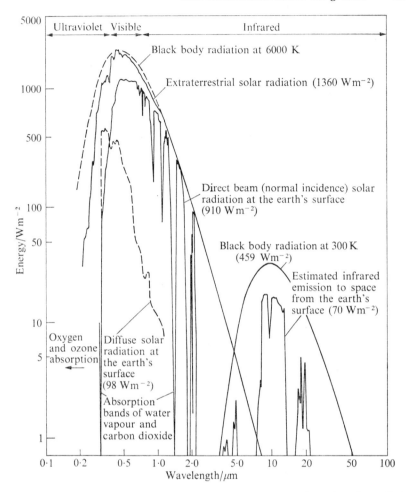

FIG. 9.1. Solar and terrestrial radiation spectra with representative values for radiation fluxes (W m⁻²). After Sellers (1965).

The rate at which energy is radiated from a body surface, the so-called radiant flux (ϕ_R W m⁻²) is proportional to the fourth power of the absolute temperature:

$$\phi_R = \epsilon\sigma T^4, \tag{9.2}$$

where σ, the *Stefan–Boltzmann constant* $= 5{\cdot}67 \times 10^{-8}$ W m⁻² K⁻⁴. The factor ϵ, known as the *emissivity*, varies with the nature of the body, and the concept of the black body has been introduced to define the perfect radiator for which $\epsilon = 1$. For all other bodies, $\epsilon < 1$, though as we shall see later, many natural bodies behave almost like black bodies at certain wavelengths.

10

The radiation incident on the surface of a body (the *irradiance*) decreases inversely as the square of the distance from the source; it also decreases with increasing inclination of the surface away from the direction normal to the source. As should be clear from Fig. 9.2, if the angle of a plane surface to the normal is α, the irradiance of the surface will be proportional to cos α. It follows that the same cosine law also applies to radiation from surfaces in a direction different to that normal to the surface.

Radiation falling on a body surface may be reflected, absorbed, or transmitted through the body.

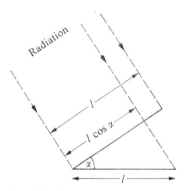

FIG. 9.2. The cosine law governing the irradiance of a plane surface inclined at an angle (α) to the plane normal to the rays.

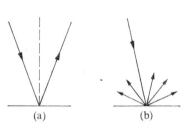

FIG. 9.3. Specular (a) and diffuse reflection (b) from a surface.

The fraction reflected (reflectivity or *reflectance*) depends on the wavelength, on the nature of the surface and on its inclination to the direction of the incoming radiation. It is important to distinguish between the reflectance at a particular wavelength (ρ_λ) and the *reflection coefficient* (ρ) corresponding to the average reflectance over a given waveband, e.g. over the whole of the visible spectrum (often called the *albedo*). Reflection may vary from the mirror or specular type in which incident and reflected beams make the same angle with the surface, to the diffuse type in which radiation is reflected at all angles, as for example from a ground glass screen (Fig. 9.3). Most natural surfaces behave as diffuse reflectors if the angle of incidence of the incident beam is less than 60 to 70° from the normal, but more like mirrors when the angle approaches 90°, the so-called *grazing incidence*.

The absorption of radiant energy (either used photochemically or transformed into heat and therefore accompanied by a rise in temperature) again varies with the nature of the body and wavelength. According to *Kirchhoff's law*, the fraction absorbed by a body at a given wavelength, the *absorptance* (α) is the same as its emissivity (ϵ) at the same wavelength. Thus a black body for which $\epsilon = 1$ at all wavelengths, will absorb all incident radiation; none will be reflected.

Besides air and water, most natural media transmit some radiation, the amount again depending on the nature of the medium, the wavelength, and the depth. For any homogeneous medium and monochromatic radiation the intensity I_x (J m^{-2}) at a depth x is given by Beer's law:

$$I_x = I_0 e^{-ax} \tag{9.3}$$

where I_0 (J m^{-2}) is the intensity at $x = 0$ and a, the so-called *absorption coefficient* is a constant, characteristic of the medium. Although strictly speaking, this law applies only to a specific wavelength, it is often used as an empirical approximation for the attenuation of solar energy by the atmosphere over the whole of the visible band. The ratio I_x/I_0 defines the *transmittance* of the medium and ax the optical density.

Having covered most of the underlying principles, we can now turn to particular examples and consider some of the important quantitative aspects.

Solar radiation

Directly or indirectly, the sun is the source of almost all the heat energy at the earth's surface. According to the most recent measurements, the radiation income on a plane surface normal to the sun's rays at the top of the atmosphere amounts to 1360 W m^{-2}. This is known as the *solar constant*. Since the sun behaves rather like a black body, its spectral range will be as illustrated in Fig. 9.1.

In its passage through the atmosphere some of the solar radiation is absorbed and some reflected by gas molecules and particles (especially liquid droplets in clouds), so that the amount and quality of the radiation reaching the earth's surface is different from that outside the atmosphere. Almost all the u.v. is absorbed by ozone, whilst water vapour and CO_2 absorb at specific wavelengths (Fig. 9.1); except for a rather intense water-vapour absorption band between 0·4 and 0·5 μm (blue light), the atmosphere is almost transparent to visible light, but there is a considerable absorption in the i.r. beyond 1 μm, again largely due to water vapour. The nearer the sun to the horizon, the longer the pathway through the atmosphere and the greater the amount of absorption (eqn 9.3). Since most absorption occurs in the blue region of the visible band this explains why the light is redder at dawn and at sunset.

With increasing inclination of the surface away from the sun, the amount of direct radiation received decreases (cosine law). Besides direct radiation, a certain amount of radiation is received indirectly from the sky because of scattering by particles and clouds. This is known as *sky light* or *diffuse light* and is the sole source of light on surfaces facing away from the sun. With clear skies and the angle of the sun $>40°$ above the horizon, diffuse light amounts to between 15 and 20 per cent of the total irradiance of the earth's surface (Fig. 9.1). Since most reflection occurs at the blue end of the spectrum, a clear sky appears blue when viewed away from the sun; clouds reflect most of the visible spectrum and therefore appear white. In contrast to direct solar radiation, for which the irradiance is maximum when the surface is at right angles to the sun's rays, maximum irradiance from diffuse light comes from directly overhead. When the sun is more than about 30° above the horizon, about 45 per cent of the direct radiation appears as visible light, but this fraction increases to about 50 per cent when diffuse radiation is taken into account.

For a given latitude and time of the year, the radiation income at the top of the atmosphere can be calculated from the distance of the earth from the sun, the solar constant and the cosine law. At a latitude of 50°, it increases from a minimum of about $8 \cdot 5 \times 10^6$ J m^{-2} per day in midwinter (January) to a maximum of about 4×10^7 J m^{-2} per day in midsummer (June). The peak noon value at ground level in southeast England with clear skies is about 900 W m^{-2} and integration over the diurnal variation gives a value of $3 \cdot 3 \times 10^7$ J m^{-2} for a 16 hour day; this suggests a loss of some 20 per cent due to attenuation by the atmosphere. In more southerly latitudes, higher peak noon values (often exceeding 1000 W m^{-2}) are offset by shorter days so that the daily total is not much higher. The most important factor is cloud which absorbs i.r. strongly and in summer reduces European values by as much as 50 per cent to between 1·5 and $2 \cdot 5 \times 10^6$ J m^{-2} per day; in winter the range for most of Europe is from 10^6 to 5×10^6 J m^{-2} per day.

If the sun is at an angle β to the horizon, and the flux of direct solar energy is ϕ_R, the irradiance of a plane horizontal surface will be $\phi_R \sin \beta$ (n.b. the cosine law applies to angles to the vertical). Monteith (1973) gives formulae for the irradiance of more complex surfaces such as spheres, horizontal and vertical cylinders etc. based on the ratio of the area of shadow cast on a horizontal surface and the area projected in the direction of the beam (Appendix 17). He also discusses the much more complicated problem of the irradiance of foliage in a canopy.

The reflection coefficients for a variety of natural surfaces are summarized in Table 9.1. These are average values expressed as a percentage of the total solar radiation; the coefficients will be different for specific wavelengths and angles of incidence.

Turning to absorptance and transmittance, it would be expected that liquid water would absorb far more intensely than water vapour; an 11 mm depth of water will absorb about 10 per cent of the incident SW radiation at 0·6 μm. Even soil transmits a little light depending on particle size; quartz sand of particle diameter 0·2 to 0·5 mm will absorb 95 per cent of the light in a depth of 1–2 mm, whereas with a particle diameter of 4–6 μm, the equivalent depth is 10 mm (Geiger 1965).

TABLE 9.1

Average reflection coefficients for natural surfaces

	% solar radiation		
Leaves	28 to 34	Man	
Crops		Eurasian	35
grass	24	Negroid	18
other crops	15 to 26	Cattle	11 to 51
Heath	14	Sheep, weathered fleece	26
Bracken	24	newly shorn	42
Tropical rain forest	13	Birds, wings	15 to 30
Deciduous woodland	18	breast	34 to 40
Conifer woodland	16		

Plane water surface, solar elevation	$> 40°$	5	
	solar elevation	$< 40°$	5 to 30
Fresh snow			80 to 85
Soil			10
Desert			30

Data from Monteith (1973).

As illustrated in Fig. 9.4, the photosynthetic pigments of leaves absorb strongly in the blue (0·40 to 0·47 μm) and red (0·60 to 0·70 μm); the relatively small absorption between 0·7 and 1·2 μm helps in reducing the heat load on leaves exposed to high levels of solar radiation. Beyond 1·2 μm radiation is almost completely absorbed by water in the tissues. Spectrophotometric measurements on the leaves of a variety of species over the range 0·4 to 2 μm (Gates *et al.*, 1965) gave mean total absorptance values (i.e. incident minus reflected minus transmitted radiation) ranging from 57 to almost 90% of the incident radiation (both direct and diffuse); the highest values refer to conifer needles which transmit negligible light. Leaf reflection coefficients quoted by Birkebak (1966) for various tree species range from 24 to 30 per cent (the lower surfaces being about 10 to 20 per cent higher than the upper surfaces) and transmittances from 17 to 28 per cent of the incident radiation. The absorptances of forest canopies are much higher than those of single leaves and for broadleaved species may amount to 85 per cent or more.

The relation between reflection, absorption and transmission of solar radiation by animal coats is rather complex because some of the incident radiation is scattered by hairs towards the skin besides being reflected away from the body. As a result, less radiation may reach the skin below a dark than light coat; at the same time a dark coat absorbs more because of its greater pigmentation (melanin) and the resulting higher temperature will mean more heat lost by thermal radiation. How far radiation penetrates the skin depends on the amount of pigmentation; it ranges from about a few tenths of a millimetre in negroid to several millimetres in light-skinned subjects.

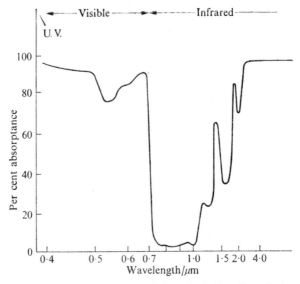

Fig. 9.4. Typical absorptance curve for a leaf of medium thickness.

Long-wave radiation

As already explained, LW radiation from bodies at normal biological temperatures occurs at wavelengths of about 10 μm with a spectral distribution as illustrated in Fig. 9.1 and at a rate of about 450 W m⁻² (assuming $\epsilon = 1$). The LW emissivities of various natural surfaces are summarized in Table 9.2; the generally high values indicate correspondingly high values for absorption.

Although the atmosphere absorbs some SW radiation, it absorbs much more LW mainly because of water vapour (in the regions 5·5 to 7·7 μm and

TABLE 9.2
Long wave (thermal) emissivities of natural surfaces
(Percent)

Leaves		Water	93 to 96
Oak	91 to 95	Fresh snow	82 to 99
Maize	94 to 95	Ice	96
others	96 to 99	Sand, dry, light	89 to 90
Animals		Sand, wet	95
Grey wolf	99	Moist bare ground	95 to 98
Caribou	100	Dry ploughed ground	90
Man	98	Desert	90 to 91
Forest	90		
Grass	90		

Data from Sellers (1965), Monteith (1973).

beyond 20 μm), CO_2 (13·1 to 16·9 μm) and clouds (all wavelengths). It is of interest that relatively little is absorbed between 8·5 and 11 μm, the so-called *atmospheric window*, corresponding to about 10 per cent of the radiation from bodies at normal temperatures. In other words, about 10 per cent of the thermal radiation from the surface of the earth is lost directly into space; the remaining 90 per cent is absorbed by the atmosphere and re-radiated, partly into space and partly back to the ground surface (as back or counter radiation). The evaluation of the LW emissivity of the atmosphere is complicated by its dependence on the water vapour content. It turns out that the emissivity is almost linearly related to the air temperature θ_a, which is closely correlated with the water vapour content. For clear skies, a useful empirical expression suggested by Monteith (1973) for calculating LW radiation from the sky (L_D) when the air temperature lies between −5 and 25 °C is:

$$L_D = 208 + 6\theta_a \text{ (W m}^{-2}) \text{ (n.b. } \theta_a \text{ in °C).} \tag{9.4}$$

When clouds are present the above expression no longer applies and allowance has then to be made for extra incoming radiation in the 8 to 13 μm band. An empirical expression due to Brunt for the apparent emissivity of a clear sky ϵ_{a0} as a function of its vapour pressure e (mb) (allowing for some variation in the constants with climate) is:

$$\epsilon_{a0} = 0.53 + 0.06\sqrt{(e)}. \tag{9.5}$$

This can be combined with the following expression for the apparent emissivity of the sky ϵ_{ac} when a fraction c is covered by cloud:

$$\epsilon_{ac} = \epsilon_{a0}(1 + nc^2), \tag{9.6}$$

where n varies with cloud type from 0·2 for low cloud to 0·04 for cirrus.

Given the apparent emissivity, L_D for a cloudy sky can then be calculated from eqn (9.2) i.e. from $\epsilon_{ac}\sigma T^4$ where T is the absolute temperature of the air.

Other empirical expressions have been used incorporating hours of bright sunshine (see p. 159).

Within the temperature range −5 to 25 °C, LW emission (L_U) from a black body ($\epsilon = 1$) at a temperature of θ °C can be calculated from the empirical relation,
$$L_U = 315 + 5.0\theta \text{ (W m}^{-2}) \text{ [}\theta \text{ in °C]} \tag{9.7}$$

(at 20 °C, this gives 415 W m^{-2} as against 420 W m^{-2} calculated from eqn (9.2)). For any other surface with an emissivity ϵ, the LW emission will be ϵL_U.

Net radiation

Having dealt with the basic principles we can now construct a simple equation representing the radiation balance for an area of the earth's surface or of organisms or communities, in terms of the net income of SW and LW radiation, the so-called *net radiation*.

First we must calculate the net income of SW solar radiation S_N. If the SW irradiance of the surface is E_s and a fraction ρ is reflected, then

$$S_N = E_s(1-\rho)$$

if the body is opaque, and $S_N = E_s(1-\rho)(1-\tau)$ if like a leaf, a fraction τ is transmitted.

If the LW radiation income from the environment is L_D and the LW radiation from the (body) surface is L_U, the net gain in all radiation, the net radiation R_N is given by:

$$R_N = S_N + L_D - L_U = E_s(1-\rho)(1-\tau) + L_D - L_U. \tag{9.8}$$

Although, it is possible to calculate R_N, it can also be measured directly for a given area of ground by means of a net radiometer (Appendix 18).

Since the general approach is the same for all systems, we shall illustrate the procedure for calculating R_N for only one system, namely that corresponding to a horizontal leaf growing over a lawn. Details for other systems are to be found in several texts including Lowry (1969) and Monteith (1973).

Suppose as in Fig. 9.5, the direct solar irradiance of a horizontal surface = 800 W m⁻² and the diffuse radiation = 100 W m⁻², giving a total SW irradiance $E_s = 900$ W m⁻². Of this a certain fraction is reflected and a certain fraction transmitted. Instead of dealing with these separately, we can use a value for the total absorptance (p. 135), say 75 per cent for an 'average'

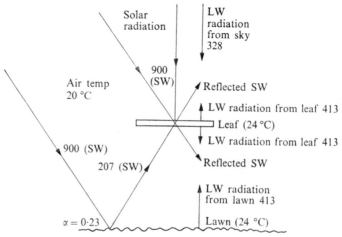

FIG. 9.5. Radiation balance of a horizontal leaf over a lawn. Figures refer to short wave (solar) and long wave (thermal) radiation fluxes in W m⁻² horizontal surface.

leaf. Hence the rate at which SW energy is absorbed $= 0.75 \times 900 = 675$ W m^{-2}.

The lower surface of the leaf will also receive SW radiation reflected from the lawn which receives the same radiation as the upper surface of the leaf, i.e. 900 W m^{-2}. If the reflection coefficient of the lawn $= 0.23$, the irradiance of the lower surface of the leaf will be $0.23 \times 900 = 207$ W m^{-2} and of this 75 per cent is absorbed by the leaf $= 159$ W m^{-2}.

The total SW radiation load on the leaf $= 675 + 159 = 834$ W per 2 m^2 (since both leaf surfaces are involved).

The upper surface of the leaf receives LW radiation from the sky and the lower surface LW radiation from the lawn.

If the sky is cloudless and the mean air temperature is 20 °C, then according to eqn (9.4) the LW irradiance of the upper surface $= 208 + 6 \times 20 = 328$ W m^{-2}.

If the temperature of the lawn is 24 °C and its LW emissivity is 0·95, then according to eqn (9.7), the LW irradiance of the lower surface $=$ $0.95(315 + 5 \times 24) = 413$ W m^{-2}.

If the leaf has the same temperature (24 °C) and emissivity (0·95) as the lawn, it will radiate at the same rate from both surfaces $= 826$ W per 2 m^2.

The net LW irradiance of the leaf (per 2 m^2) $= 328 + 413$ W and assuming an absorptance of 0·95 (the same as its emissivity) the net LW load per 2 m$^2 = 0.95 \times 741 = 704$ W.

The net *loss* in LW energy is therefore $704 - 826 = -122$ W per 2 m^2.

The total radiation absorbed per 2 m^2 surface

$$= 834 \text{ (SW)} - 122 \text{ (LW)} = 712 \text{ W}.$$

Hence the net radiation $R_N = 356$ W m^{-2}.

Heat balance

Although the above example is by no means typical of all systems or at all times, (e.g. during the night, R_N for the ground surface is usually negative) it is clear that we are dealing with an instantaneous rather than an equilibrium situation. As the leaf gains energy, its temperature will rise and its emission will increase until an equilibrium is eventually established. If the equilibrium temperature of the leaf is θ_1 °C, then its LW emission per 2 m^2

$$[= 2 \times 0.95(315 + 5\theta_1)]$$

must balance the radiative load $[= 834 \text{ (SW)} + 704 \text{ (LW)}]$ according to which $\theta_1 \approx 100$ °C. This of course is an entirely spurious situation since as the leaf temperature increases, more and more heat will be lost by convection (either free or forced, depending on wind conditions). For example, with a leaf say, 0·1 m long in the direction of wind moving at a speed of U m s^{-1}, it can easily be shown from the eqn (Table 7.4) that with laminar flow over *both surfaces*, the convective heat loss for a temperature difference of say, 4 °C between the leaf and air, $\approx 101\sqrt{(U)}$ W m^{-2}. To balance the net radiation of 356 W m^{-2} therefore, U would have to be just over 10 m s^{-1} (≈ 4 miles h^{-1}). At lower wind speeds, the leaf temperature could only be maintained at 24 °C if the extra heat was dissipated by evaporative cooling (transpiration).

We can extend this argument to the heat balance of other organisms and to areas of land over various time intervals in the form of a general heat balance:

$$R_N + M = G + H + LE, \tag{9.9}$$

where M represents the net gain of heat energy due to metabolic activity. In the case of plants it is a negative quantity corresponding to the use of some of the light energy for photosynthesis. According to Monteith (1973) maximum rates, in terms of the energy content of the photosynthetic products, range from about 7 to 16 W m^{-2}. During the day this is negligible compared with radiation incomes of 300 to 500 W m^{-2}. At night a certain amount of heat is produced by respiration, of the order of 3 W m^{-2} for a crop, and whilst this may have little influence on LW radiation losses during a clear night, it may be of the same order as these losses during cloudy conditions.

In the case of animals, heat energy released by metabolic activity must be added to R_N. The basal metabolic rate corresponding to resting subjects in a thermo-neutral environment (when metabolism is independent of the external temperature) can be expressed as a function of body weight or surface area. Over a range of species it amounts to an average of about 50 W m^{-2} body surface. Again this is usually small compared with the radiation income during the day in the open, but may not be so in shade; furthermore, muscular activity may increase the heat output by as much as 10 times.

G represents the change in heat stored in the body and/or in the ground. The amount of heat stored in vegetation generally amounts to only a few per cent of the radiation income; for homeotherms, with a constant body temperature, the storage term can also be ignored. In the case of areas of ground however, heat conducted into and stored in the soil may represent an appreciable fraction of the radiation income during the day (see p. 84). During the night almost the whole of the LW radiation from the ground surface is derived from heat stored in the soil. Under a cover of vegetation the radiation income to the soil is reduced and the transfer of heat to the ground may then amount to only about 5 to 10 per cent of the net radiation during the day. Over short intervals, G may be calculated from the temperature gradient in the soil or measured directly by a soil heat flux plate. Over longer periods, provided that they begin and end at the same time of day, it may be assumed in practice that there has been no net change in heat so that $G = 0$.

H (previously denoted as ϕ_h) represents the loss of sensible heat mainly by convection to the air and LE the loss of (latent) heat corresponding to an evaporation rate E.

As we shall be dealing mainly with areas of ground covered by vegetation, the heat balance may be expressed more or less completely by:

$$R_N - G = H + LE. \tag{9.10}$$

It follows that if $R_N - G$ is known, either by calculation or direct measurement, and if we can determine H or the ratio H/LE, we can evaluate LE and hence the rate of evaporation from the vegetation and ground. This ratio H/LE is known as the Bowen ratio (β) and from eqn (9.10) it can be shown that:

$$LE = \frac{R_N - G}{1 + H/LE} = \frac{R_N - G}{1 + \beta}. \tag{9.11}$$

The application of the heat balance and the Bowen ratio to the measurement of evaporation from crops will be discussed later.

Turbulent transport

When air moves over the ground a turbulent boundary layer is formed characterized by rapid changes in wind direction and velocity due to the random movement of eddies. As explained earlier (p. 12) these eddies can be regarded as 'parcels' of air moving in all directions, so providing the main mechanism for the transport and mixing of the various properties of the air such as its heat, water vapour and CO_2 content. Since just as many eddies move upwards as downwards, then if the air is warmer or contains more water vapour near the surface than higher up, (i.e. there is a gradient in these properties with height), upward moving eddies will carry with them more heat and water vapour than downward moving eddies. The net effect is a vertical transport of these properties equivalent to a kind of convective transport known as *turbulent* or *eddy diffusion*.

The importance of turbulent diffusion cannot be overemphasised since if it did not occur, life as we know it could not exist. Although molecular diffusion of the air properties still takes place, it is far too slow (of the order of 10^3 to 10^4 times slower than turbulent diffusion) and could never supply or remove O_2 and CO_2 at rates sufficient to maintain essential life processes, nor would heat exchange occur fast enough to prevent enormous diurnal changes in temperature at the earth's surface which would be lethal to most organisms.

Because of the importance of turbulent diffusion, considerable attention has been paid to the factors determining its rate for the various air properties concerned, and to its measurement in field conditions. Before we can discuss these usefully, we need to discuss in rather more detail the nature of the turbulent boundary layer.

The turbulent boundary layer over the ground

If the surface of the ground was perfectly smooth, the effect of wind blowing over it would not be unlike the development of a turbulent boundary layer over a smooth flat plate at high (*Re*), with a laminar sub-layer between the surface and the turbulent zone (Chapter 4). The earth's surface however is characteristically rough, especially when covered with vegetation; under these circumstances there is little or no laminar sub-layer and turbulence

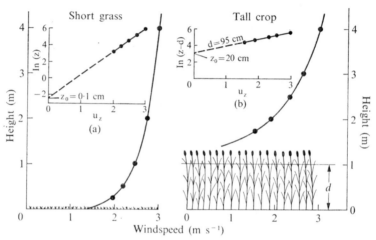

FIG. 9.6. Wind profiles over short grass (a) and a tall crop (b) when the windspeed at 4 m above the ground is 3 m s^{-1}. The filled circles represent hypothetical measurements from sets of anemometers *Insets:* plots of mean windspeed (u_z) against log height (z) or log ($z-d$). z_0 = roughness length, d = zero plane displacement. z, z_0, and d in cm. From Monteith (1973).

extends practically to the surface, being maintained in this region by the disturbance imposed on air movement through the drag of the roughness elements. The rougher the surface, the greater will be the drag and the deeper the layer of air affected. The situation is illustrated by the wind profiles over a short grass cover and over a tall crop in Fig. 9.6.

Measurements on wind profiles over bare ground or *short* grass covers (Fig. 6a) have shown that if the temperature gradient in the air is small, the mean horizontal wind velocity (u_z) is directly proportional to the logarithm of the height (z) above the surface. However, when u_z is plotted against log z (*inset*, Fig. 6a), the straight line extrapolated back to $u_z = 0$, cuts the log z axis at a certain height, z_0 so that the relationship can be expressed as:

$$u_z \propto (\log z - \log z_0) \qquad or, \qquad u_z \propto \log\left(\frac{z}{z_0}\right). \qquad (9.12)$$

The height z_0 above the ground surface, at which the windspeed is assumed to be zero, is known as the *roughness length*. More will be said about this later.

Although the logarithmic nature of the turbulent wind profile can be deduced theoretically, a complete solution can only be obtained experimentally and to simplify matters, we might refer back to Fig. 2.4 for turbulent flow in pipes. These data may also be applied to turbulent flow over flat surfaces. By incorporating a friction velocity $u_* = \sqrt{(\tau_0/\rho)}$ (where $\tau_0 =$ surface shear stress and $\rho =$ air density) it was shown that beyond a certain distance from the surface, the plot of u/u_* against log (zu_*/ν) could be fitted by the equation:

$$\frac{u_z}{u_*} = 5\cdot75\log_{10}\left(\frac{u_* z}{\nu}\right) + 5\cdot5 \quad or \quad \frac{u_z}{u_*} = 2\cdot5\ln\left(\frac{u_* z}{\nu}\right) + 5\cdot5,$$

where $u_z =$ mean velocity at a distance z from the surface.

This may be written as:

$$\frac{u_z}{u_*} = 2\cdot5\ln(z) + 2\cdot5\ln\left(\frac{u_*}{\nu}\right) + 5\cdot5 \tag{9.13}$$

If as before, we stipulate that $u_z = 0$ when $z = z_0$, we obtain a quantitative solution identical in form with eqn (9.12) which agrees with that obtained by other approaches:

$$u_z = 2\cdot5u_*\ln\left(\frac{z}{z_0}\right). \tag{9.14}$$

It should be noted that the wind profile can be represented by other expressions, but the one given above is the most commonly used.

In the case of tall crops (see Fig. 9.6b), for which the velocity profile is displaced still further up the z axis, the logarithmic relationship can only be maintained if another length, d known as the *zero plane displacement length* is subtracted from z (*inset*, Fig. 6b), giving eqn (9.14) in the form:

$$u_z = 2\cdot5u_*\ln\left(\frac{z-d}{z_0}\right). \tag{9.15}$$

(n.b.: most texts quote eqns (9.14) and (9.15) with a factor $1/k$ instead of $2\cdot5$; k is known as the *von Karman constant* and as we see $\approx 0\cdot4$.)

It should now be clear that the roughness length z_0 is such as to make the velocity (and therefore the momentum) of the wind zero at a height $z_0 = z-d$; this has the advantage of providing a reference level for zero momentum, often described as a 'sink' for momentum. Nevertheless, z_0 has a certain physical significance in so far as it provides a quantitative measure of the size, shape, and spacement of the roughness elements.

Measurements have shown that z_0 may be related empirically to the height h of the roughness elements (e.g. plant height) by:

$$\log_{10}z_0 = \log_{10}h - 0.98 \quad \text{or approximately,} \quad z_0 = 0.13h. \qquad (9.16)$$

Characteristic values for z_0 range from about 10^{-5} m for very smooth surfaces (mud flats, ice) to between 0.05 and 0.3 m for tall grass and vegetable crops and up to several metres for trees, (see Table 9.3). Depending on the extent to which the vegetation bends in the wind, z_0 generally decreases with increasing wind speed. Szeicz, Endrodi, and Tajchman (1969) found that as the wind speed increased from 0 to 4 m s^{-1}, z_0 decreased from about 6 to 2 m for pine forest, from 0.09 to 0.04 m for a potato crop and from 0.08 to 0.03 m for lucerne; a similar decrease is shown by the grass data in Table 9.3.

The zero plane displacement, d represents the level above which the wind velocity can be assumed to be proportional to log z. Like z_0 it must be determined experimentally (e.g. by fitting a value to a measured velocity profile so as to give the log relationship), but as a rough approximation, it can be estimated from h, the crop height:

$$d = 0.63h. \qquad (9.17)$$

For the most part d is independent of wind speed and published values range from 0 for very smooth surfaces to between 0.2 and 0.3 m for tall grass and vegetable crops, and several metres for forests.

The values for z_0 quoted from Sellers (1965) in Table 9.3 have been recomputed from the original velocity profiles to include d and so allow for the application of eqn (9.14) instead of (9.15).

Eqns (9.14) and (9.15) have been very widely used in field measurements of turbulent diffusion and it is important to consider the conditions under which they apply. For example, they require a surface of uniform roughness (e.g. plants of reasonably uniform height and spacement) and a boundary layer characteristic of that surface. When the roughness of the surface changes, as for example from open field to forest, the new profile develops only slowly downwind. As a rough rule, at a distance x downwind from a change in surface roughness (so-called *fetch*), the depth z of profile characterising the new surface is given by $z = x/40$. In other words, if the fetch over the new surface is 100 m, the depth of the fully developed profile will be 2.5 m. Furthermore, as we shall see later, many measurements are based on the assumption of a constant flux of momentum and this can only be assured in the lowest 15 per cent or so of the boundary layer where turbulence is entirely maintained by the drag of the surface roughness elements. It turns out that for every metre of boundary layer to which this applies, a fetch of some 200 metres is required.

Perhaps the most important restriction however, is the need for the temperature gradient in the air to be small so that density changes in the air do not

TABLE 9.3

Some aerodynamic properties of crops

	Crop height m	Wind speed u(m s^{-1})	at height z(m)	z_0 m	d m	u_* m s^{-1}	τ N m^{-2}	K_m m^2 s^{-1}	C_D
Short grass[a]	0·1	3	4	10^{-3}	7×10^{-3}	0·15	0·027	0·25 at 4 m	0·005 at 4 m
Tall crop[a]	1	3	4	0·2	0·95	0·46	0·25	0·58 at 4 m	0·045 at 4 m
Tall grass[b]	0·6	1·48	2	0·154	—	0·23	0·06	0·18 at 2 m	0·048 at 2 m
	0·6	3·43	2	0·114	—	0·48	0·28	0·38 at 2 m	0·039 at 2 m
	0·6	6·22	2	0·080	—	0·77	0·72	0·62 at 2 m	0·031 at 2 m

[a] From Monteith (1973); [b] from Sellers (1965).

lead to complicating buoyancy forces. This leads us to the factors influencing air stability.

Air stability and lapse rates

Imagine a 'bubble' of air near ground level with the same temperature as the surrounding air, and suppose that we move this bubble to a greater height (Fig. 9.7). As the air pressure decreases the bubble will expand and if

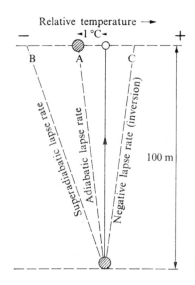

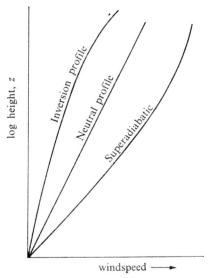

FIG. 9.7. Lapse rates. Pecked lines indicate temperature profiles in the bulk air. For explanation see text.

FIG. 9.8. Effect of air stability on the wind velocity profile.

there is no heat exchange with the surrounding air (so-called *adiabatic expansion*) it will cool at a rate of about 1 °C per 100 m rise in dry air. This is known as the *dry adiabatic lapse rate* (lapse rate refers to the fall in temperature with altitude).

If the temperature gradient in the surrounding air corresponds exactly to this lapse rate, then whatever its height, the bubble will be at the same temperature as the surrounding air (Fig. 9.7A); it will therefore have the same density and will neither rise nor fall. We then have what are known as *neutral conditions of air stability*.

If on a sunny day, the ground surface warms up and produces a steeper temperature gradient in the air, a so-called *super-adiabatic lapse rate* (Fig. 9.7B), then the bubble of air which starts out near ground level with the same temperature as the surrounding air, will always be warmer than the surrounding air when raised to a greater height; being less dense, therefore, it will continue to rise by buoyancy forces. The condition of the air is now said to be *unstable* in so far as upward rising air masses will never be in equilibrium with the surrounding air.

Conversely, if the lapse rate is less steep than adiabatic, the air bubble at greater heights will be cooler and denser than the surrounding air so that it will tend to sink down again. This situation often occurs during clear cold nights when, because of excessive radiation, the ground temperature falls below that of the air so that the lapse rate is actually reversed (Fig. 9.7C). Such conditions are known as *inversions*, and the fact that upward movement of the air is suppressed leads to very stable conditions with little if any mixing. Hence the accumulation of pollutant fumes (e.g. smog) near the ground when inversions occur.

If the air is moist, the cooling of a rising air mass due to expansion is offset by the condensation of water vapour and the liberation of latent heat. The fall in temperature with height is then less than in dry air and decreases with increasing air temperature from about 0·5 to 0·3 °C per 100 m (the *moist adiabatic lapse rate*). Moist air rising through dry air will therefore lead to instability since it will always be warmer than the surrounding air.

The velocity profile eqn (9.15) refers strictly to *neutral* conditions; generally speaking, these occur only for short periods, usually after sunrise and before sunset. At other times of the day, when the surface of the ground is very warm, super-adiabatic conditions often prevail, whilst during the night, inversions are common. The effect of these conditions on the velocity profile is illustrated in Fig. 9.8. A super-adiabatic lapse rate promotes turbulence, and the more rapid downward transport of momentum from the freely moving air streams higher up produces a more uniform velocity profile, except near the surface. By contrast, inversions depress turbulence so that there is relatively little vertical transport of momentum with the result that the velocity near the ground is lower and the profile more sharply curved.

Expressions have been proposed which make allowance for the effect of buoyancy on the velocity profiles, but we shall not go into these. We shall turn instead to some quantitative aspects of turbulent diffusion and especially how it may be measured in the field.

The measurement of turbulent diffusion

We have seen in the previous two chapters how estimates can be made of convective heat and mass transport from individual organs and organisms under conditions of turbulent flow. Unfortunately these formulae cannot be applied to relatively large areas of ground surface so that other approaches have to be adopted. Before we can consider these however, we must look again at the equations for the molecular diffusion of heat and mass (7.2; 8.1). Since we are now concerned with vertical diffusion (i.e. perpendicular to the ground) we shall follow convention and use the z co-ordinate.

11

For mass diffusion we have $\phi_c = -D(dc/dz)$

and for heat conduction, $\phi_h = -\lambda(d\theta/dz)$.

It will be noted that the units of the mass diffusion coefficient $D(m^2\,s^{-1})$ are different from those of the thermal conductivity λ $(J\,m^{-1}\,s^{-1}\,K^{-1})$. The two equations can be made formally consistent by replacing λ by the thermal diffusivity $\alpha(= \lambda/\rho c_p)$ which also has units of $m^2\,s^{-1}$:

$$\phi_h = -\alpha\frac{d(\rho c_p\theta)}{dz}. \tag{9.18}$$

In other words we can regard heat transfer as occurring along a gradient of $\rho c_p\theta$ i.e. a gradient of the heat *content* per unit volume of air, corresponding to that of mass transport along a gradient of mass content per unit volume.

The transport of momentum can also be regarded as a diffusion process. Since the momentum of the molecules is directly proportional to their speed u, the velocity gradient du/dz within the boundary layer corresponds to a momentum gradient so that by analogy with heat and mass transport, this can be regarded as the 'driving force' for the molecular diffusion of momentum. We can therefore look upon eqn (1.1) $\tau = \eta\,du/dz$ as being analogous to a diffusion equation with the shear stress τ now interpreted as a flux of momentum. This is made clearer if we replace η by ρv to give a coefficient with the same units as D and α:

$$\tau = v\frac{d(\rho u)}{dz}, \tag{9.19}$$

τ can now be seen to correspond to a flux of momentum (mass$\times$velocity per unit area per unit time $= kg\,m\,s^{-1}\,m^{-2}\,s^{-1} = kg\,m^{-1}\,s^{-2}$) along a gradient of ρu i.e. momentum per unit volume.

Since the same gradients are involved, we can write the turbulent diffusion equations in the same form as those for molecular diffusion, replacing the molecular diffusion coefficients by the corresponding turbulent diffusion coefficients K:

For momentum, $\tau\,(kg\,m^{-1}\,s^{-2}) = K_m\dfrac{d(\rho u)}{dz}$;   (9.20a)

For heat, $\phi_h = H(W\,m^{-2}) = -K_h\dfrac{d(\rho c_p\theta)}{dz}$;   (9.20b)

For mass, $\phi_c\,(kg\,m^{-2}\,s^{-1}) = -K_c\dfrac{dc}{dz}$.   (9.20c)

Note that the equations for heat and mass still retain the negative sign because these quantities are assumed to decrease with increasing distance z from the surface, whereas like the velocity, the momentum increases with z. The term ϕ_h has been replaced by the more conventional symbol H.

As far as mass transport is concerned, we shall only be concerned with water vapour and CO_2. Water vapour concentrations are usually expressed in terms of the specific humidity q (kg per kg moist air) or more commonly as the vapour pressure e (mb). For the present we shall use $q(= c/\rho)$ where ρ is the air density) (Appendix 15). Following convention, we denote the evaporation rate ϕ_E as E and the corresponding turbulent diffusion coefficient as K_w. Eqn (9.20c) now becomes:

$$\text{For evaporation, } E(\text{kg m}^{-2}\,\text{s}^{-1}) = -K_w \frac{d(\rho q)}{dz}. \qquad (9.21)$$

Although the concentration of CO_2 in the air is usually expressed as a volume per unit volume of air, to avoid confusion we shall continue with the mass concentration c (kg m^{-3}), replacing the flux by C and the turbulent diffusion coefficient by K_c:

$$\text{for } CO_2, \quad C(\text{kg m}^{-2}\,\text{s}^{-1}) = -K_c \frac{dc}{dz}. \qquad (9.22)$$

n.b. in the case of active photosynthesis, the negative sign is deleted because the CO_2 concentration then increases with height, z.

In practice the evaluation of the various fluxes is usually based on measurements of the corresponding properties at different heights above the ground surface or crop (Fig. 9.9). As in the case of molecular diffusion, the appropriate working equations are obtained by integrating the above equations between the given heights. However, unlike their molecular counterparts, the turbulent diffusion coefficients are not constant, but vary with height and with air stability (as would be expected from the movement of the eddies). To overcome this, a new coefficient $\bar{K}$ is defined which represents the average value of K between the given heights under the given conditions, so that the working equations for transport between heights (1) and (2) become:

$$\tau = \rho\bar{K}_m(u_2 - u_1) \quad \text{kg m}^{-1}\,\text{s}^{-2} \qquad (9.23a)$$

$$H = \rho c_p \bar{K}_h(\theta_1 - \theta_2) \quad \text{W m}^{-2} \qquad (9.23b)$$

$$E = \rho\bar{K}_w(q_1 - q_2) \quad \text{kg m}^{-2}\,\text{s}^{-1} \qquad (9.23c)$$

$$C = \bar{K}_c(c_1 - c_2) \quad \text{kg m}^{-2}\,\text{s}^{-1} \qquad (9.23d)$$

where u_1, u_2 (m s^{-1}); θ_1, θ_2 (°C); q_1, q_2 (kg kg^{-1}) and c_1, c_2 (kg m^{-3}) refer respectively to the values of these properties at heights z_1 and z_2. Since we are dealing with turbulent conditions, these values will be the means of measurements made over a period of about 30 to 60 minutes (p. 13).

The above measurements do not in themselves allow for the direct evaluation of the fluxes since the coefficients are still unknown. Considerable research has gone into devising ways and means to deal with this problem, but we shall only consider a few of the more popular techniques. Further details can be obtained from several texts and reviews including Webb (1965) and WMO (1968); details of the instrumentation which constitutes a vital part of the research are given, among others, by Sěsták, Catsky and Jarvis (1971) and Monteith (1972).

A common approach to the problem is to assume that the turbulent diffusion coefficients are equal and to use the ratios, e.g. if $\bar{K}_m = \bar{K}_h = \bar{K}_w$ then:

$$\frac{E}{\tau} = \frac{(q_1-q_2)}{(u_2-u_1)}; \qquad \frac{H}{E} = \frac{c_p(\theta_1-\theta_2)}{(q_1-q_2)}. \qquad (9.24)$$

The assumption of equality rests on the argument that all properties of the air are transported by the same eddy motion and undergo the same mixing. This is not strictly true, especially between $\bar{K}_m$ on the one hand and $\bar{K}_h$ and $\bar{K}_w$ on the other; this is because momentum transfer is due largely to pressure forces acting on individual roughness elements, whilst heat and matter must cross the laminar boundary layer enveloping each element by molecular diffusion. Nevertheless, it has been argued that if the measurements are restricted to a certain height above the surface where turbulence is almost entirely brought about by drag of the roughness elements, equality can be assumed in practice. At greater heights buoyancy forces may transfer heat more readily than water vapour or momentum and the situation becomes even more doubtful when the air conditions are not neutral. A further reason for restricting profile measurements to the lower part of the boundary layer is the effect of the fetch. As a rough guide, $z-d$ should not exceed 0·05 per cent of the fetch over a uniform surface, but measurements may be made up to about 0·1 per cent under certain circumstances. In practice, this often means the restriction of measurements to 1 or 2 metres above the surface of a crop.

The aerodynamic method

In this method, use is made of the ratios incorporating momentum transport (τ) since this can be determined independently from the velocity profile.

If the mean horizontal wind velocities, u_1 and u_2 are measured at two heights, z_1 and z_2, then from eqn (9.15):

$$u_1 = 2 \cdot 5 u_* \ln\left(\frac{z_1-d}{z_0}\right) \qquad \text{and} \qquad u_2 = 2 \cdot 5 u_* \ln\left(\frac{z_2-d}{z_0}\right),$$

whence, by a little rearranging,

$$u_* = \frac{(u_2-u_1)}{2 \cdot 5 \ln\left(\frac{z_2-d}{z_1-d}\right)} \qquad (9.25)$$

Within the turbulent zone, it is assumed that the momentum flux τ is constant; therefore, $\tau = \tau_0$ and $u_*^2 = \tau_0/\rho = \tau/\rho$. From eqn (9.23a) therefore,

$$\bar{K}_m = \frac{\tau}{\rho(u_2-u_1)} = \frac{u_*^2}{(u_2-u_1)} = \frac{(u_2-u_1)}{\left[2 \cdot 5 \ln\left(\dfrac{z_2-d}{z_1-d}\right)\right]^2} \tag{9.26}$$

Assuming that the turbulent diffusion coefficients are the same, this value of $\bar{K}_m$ can be applied to the diffusion of other properties, e.g. in the case of evaporation, (9.23a–c), if $\bar{K}_w = \bar{K}_m$:

$$E = \rho \frac{(u_2-u_1)(q_1-q_2)}{\left[2 \cdot 5 \ln\left(\dfrac{z_2-d}{z_1-d}\right)\right]^2}. \tag{9.27}$$

This is the well-known Thornthwaite–Holzman equation for evaporation (usually quoted without d). As we see, it involves the measurement of the mean horizontal wind velocity (u) and air humidity (q) at two known heights above ground (z).

Combined aerodynamic and energy balance methods

A popular approach to the determination of evaporation from land surfaces makes use of the Bowen ratio (eqn 9.11). Multiplying E in eqn (9.24) by L the latent heat, and substituting this in eqn (9.11) gives:

$$\frac{H}{LE} = \frac{c_p(\theta_1-\theta_2)}{L(q_1-q_2)} \quad \text{and} \quad LE = \frac{(R_N-G)}{1+\dfrac{c_p(\theta_1-\theta_2)}{L(q_1-q_2)}}. \tag{9.28}$$

In other words, the evaporation rate E can be determined from a knowledge of the available net radiation (R_N-G) and the differences in temperature (θ) and in specific humidity (q) measured at two suitable heights above the surface.

A somewhat simpler version of the above makes use of wet and dry bulb thermometers (or the equivalent) installed at the two heights. It can then be shown (Appendix 19) that:

$$LE = (R_N-G)\left[1 - \frac{\gamma}{s+\gamma} \cdot \frac{(\theta_1-\theta_2)}{(\theta_{w1}-\theta_{w2})}\right] \tag{9.29}$$

where θ_{w1}, θ_{w2}, θ_1, θ_2 are respectively the wet and dry bulb temperatures at heights (1) and (2) above the ground, $\gamma =$ psychrometric constant and $s =$ the slope of the saturation vapour pressure–temperature curve at the mean temperature of the wet bulbs (Fig. 9.10).

The above equation forms the basis of the recently developed EPER (energy partition evaporation recorder) which appears to have been used with some success to record evaporation rates automatically (McIlroy 1971).

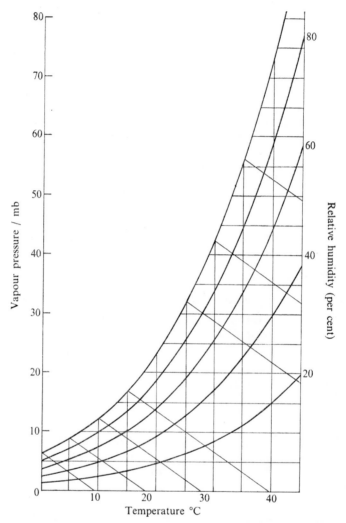

FIG. 9.10. Relationship between saturation vapour pressure of water and temperature.

Once E is known, the ratio with some other quantity may be taken to evaluate the flux of that quantity (e.g. CO_2).

Diffusion resistances

The concept of a diffusion resistance described earlier can be usefully extended to turbulent diffusion above the ground. It has proved particularly valuable in relating evaporation from crops not only to radiation and wind conditions, but also to stomatal behaviour. A similar approach can be adopted for CO_2 diffusion, which is influenced by many of the same factors.

Applying the now familar Ohm's law analogy to eqn (9.23) for turbulent diffusion (i.e. flux = driving force divided by diffusion resistance) we have:

$$\text{For momentum,} \quad \tau(\text{kg m}^{-1}\text{ s}^{-2}) = \rho.\frac{(u_2-u_1)}{r_{am}} \tag{9.30a}$$

$$\text{For heat,} \quad H(\text{W m}^{-2}) = \rho c_p.\frac{(\theta_1-\theta_2)}{r_{ah}} \tag{9.30b}$$

$$\text{For water vapour,} \quad E(\text{kg m}^{-2}\text{ s}^{-1}) = \rho.\frac{(q_1-q_2)}{r_{aw}} \tag{9.30c}$$

$$\text{or } LE(\text{W m}^{-2}) = \frac{\rho c_p}{\gamma}.\frac{(e_1-e_2)}{r_{aw}} \tag{9.30d}$$

where r_{am}, r_{ah}, r_{aw} are the corresponding diffusion resistances in air; a check on the units will confirm that these are s m^{-1} in all cases (most texts still quote r_a in terms of s cm^{-1}). Note that the eqn (9.30d) for the flux of latent heat (LE) is expressed in terms of a difference in vapour pressure (e_1-e_2) to conform with common practice; this is based on $q \approx 0.622e/p$ (Appendix 15) and γ, the psychrometric constant $= pc_p/0.622L$ (eqn 8.35).

Evaporation from open water surfaces

Suppose that instead of measuring the temperature and humidity at two levels above the surface, measurements are made at one height z giving θ_z and e_z, the other level being the surface itself, $z = 0$. The vapour pressure at the surface will be the saturation vapour pressure, e_{s0} at the temperature of the surface, θ_0. An equation based on combined aerodynamic and heat balance principles can then be constructed along the lines of eqn (9.28), but now incorporating r_{ah} and r_{aw}:

$$LE = \frac{(R_N-G)}{1+\left(\gamma\dfrac{r_{aw}}{r_{ah}}\right)\left(\dfrac{\theta_0-\theta_z}{e_{s0}-e_z}\right)}.$$

The practical drawback of this equation is that it is very difficult to measure the surface temperature of the water. As explained in Appendix 20, θ_0 can be eliminated by introducing a new constant Δ corresponding to the slope of the saturation vapour pressure-temperature curve at the air temperature θ_z (Fig. 9.10). This gives:

$$LE = \frac{(R_N-G)+(\rho c_p/r_{ah})(e_{sz}-e_z)}{\Delta+\gamma(r_{aw}/r_{ah})}. \tag{9.31}$$

If, as in practice, it is assumed that $r_{aw} = r_{ah} = r_a$, then:

$$LE = \frac{(R_N-G)+(\rho c_p/r_a)(e_{sz}-e_z)}{\Delta+\gamma} \tag{9.32}$$

where e_{sz} is the saturation vapour pressure at the air temperature θ_z and e_z the actual vapour pressure at height z. The difference, $(e_{sz}-e_z)$ corresponds to the vapour pressure deficit of the air.

The above equation was first proposed by Monteith (1965) and is similar in form to an equation originally derived by Penman (1948) using a different notation. We see that in common with other combination formulae, it contains a radiation term (R_N-G) representing the supply of latent heat energy, and an aerodynamic term (r_a) representing the degree of turbulence.

Evaporation from leaves and crop canopies

Eqn (9.32) only applies to evaporation from open water or wet surfaces at which the vapour pressure corresponds to the saturation vapour pressure at the surface temperature. With leaves, evaporation occurs from the surfaces of the mesophyll cells inside, and the leaf surface itself is dry. At the same time an extra resistance r_s is involved corresponding to the pathway from the leaf interior, through the stomatal pores, to the outer surface of the leaf (Fig. 9.11). As explained earlier, r_s will be determined by the degree of stomatal opening.

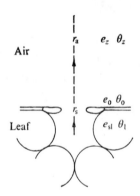

FIG. 9.11. Diffusion of water vapour from interior of a leaf through the stomatal pores into the air. r_s, r_a refer respectively to the leaf (stomatal) and air diffusion resistances, e_{sl}, e_0, e_z and θ_1, θ_0, θ_z to the vapour pressures and temperatures within the leaf, at the outer surface of the leaf, and in the air well above the leaf.

In dealing with evaporation from leaves (i.e. transpiration) we can consider the two pathways separately (mesophyll to outer surface, outer surface to bulk air) or together, with the resistances r_s and r_a in series with each other.

If as shown in Fig. 9.11 the vapour pressure inside the leaf = saturation vapour pressure e_{sl} at the leaf temperature θ_1, and the vapour pressure at the leaf surface = e_0, we have a vapour pressure difference $(e_{sl}-e_0)$ acting across a resistance r_s; applying eqn (9.30d):

$$LE = \frac{\rho c_p(e_{sl}-e_0)}{\gamma r_s}.$$ (9.33)

For the second part of the pathway, leaf surface to air, the vapour pressure difference is $(e_0 - e_z)$ and the resistance, r_{aw} so that:

$$LE = \frac{\rho c_p(e_0 - e_z)}{\gamma r_{aw}}. \qquad (9.34)$$

For the whole pathway with a vapour pressure difference $(e_{sl} - e_z)$ and a combined resistance $(r_s + r_a)$ we have:

$$LE = \frac{\rho c_p(e_{sl} - e_z)}{\gamma(r_s + r_{aw})}. \qquad (9.35)$$

Again by incorporating Δ, the slope of the SVP-temperature curve at the air temperature θ_z (Fig. 9.10), we can derive an equation formally identical with eqn (9.31) but with γ replaced by $\gamma(r_s + r_{aw})/r_{ah}$ (see Appendix 20). Assuming that $r_{aw} = r_{ah} = r_a$ gives:

$$LE = \frac{(R_N - G) + (\rho c_p/r_a)(e_{sz} - e_z)}{\Delta + \gamma(1 + r_s/r_a)} \qquad (9.36)$$

where as before $(e_{sz} - e_z)$ represents the vapour pressure deficit of the air.

Eqn (9.36) can be applied to crop canopies if r_s is replaced by r_c, a canopy resistance representing the combined diffusion resistances of all the leaves in the canopy (some authors still retain r_s for a canopy and call this a surface resistance).

$$LE = \frac{(R_N - G) + (\rho c_p/r_a)(e_{sz} - e_z)}{\Delta + \gamma(1 + r_c/r_a)}. \qquad (9.37)$$

In applying eqn (9.37) in the field, measurements of $(R_N - G)$ and e_z are relatively straightforward; the real problem lies with r_a and r_c.

Air resistance (r_a)

This may be determined experimentally from the mean horizontal wind velocity u_z at height z above ground level if d and z_0 are known and air stability conditions are satisfactory; it can then be shown (Appendix 21) that:

$$r_a(= r_{am}) = \frac{\left\{2 \cdot 5 \ln\left(\frac{z - d}{z_0}\right)\right\}^2}{u_z}. \qquad (9.38)$$

It follows from eqn (9.38) that if z_0 remains constant, the air diffusion resistance r_a should be inversely proportional to the wind velocity, u_z and that the inverse, $1/r_a$ (known as the *eddy conductivity*) should be directly proportional to u_z. This is confirmed, at least for open water and for potato and lucerne crops by the data of Szeicz *et al.* (1969) illustrated in Fig. 9.12. The anomalous behaviour of the tree crop is probably due to crown deformation with wind.

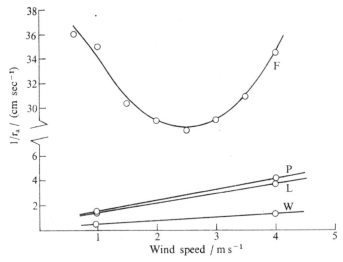

FIG. 9.12. Variation in eddy conductivity ($1/r_a$) with windspeed measured at 1·7 m above the surface for pine forest (F), lucerne (L) and potato (P) crop and open water (W). The crop height for lucerne and potato is 0·6 m. Note change in scale of ordinate for the forest. From Szeicz *et al.*, (1969). *Water Resources Res.*, **5**, 380. Copyright by American Geophysical Union.

The relatively low eddy conductivity of the open water surface compared with that of the forest can be attributed to the greater roughness of the latter. For the same absorption of radiant energy and the same vapour pressure deficit in the air therefore, evaporation of free water from the surface of the forest canopy, e.g. of rainwater intercepted by and retained on the canopy, should be correspondingly more rapid than from the same area of open water surface at ground level. This has been confirmed in field experiments on a number of occasions.

The same relation between the roughness of a vegetative cover and eddy conductivity also helps to explain, at least to some extent, why forests generally yield less water from catchment areas than more lowly vegetation such as grass. This is of considerable practical importance in water conservation. However, in comparing one vegetative cover with another, other factors have to be taken into account, one of the more important of which is the crop resistance, r_c.

Crop resistance (r_c)

As already explained, this parameter reflects the degree of stomatal control of transpiration and it may be measured in various ways depending on the equipment available and whether short or long term averages are required. Short term averages on an hourly basis or so are particularly valuable in that they provide data on diurnal fluctuations affecting evaporation and photosynthesis (i.e. CO_2 diffusion).

The *profile method*, which allows for short term measurements is based on eqn (9.33) with r_c replacing r_s:

$$r_c = \frac{\rho c_p (e_{sc} - e_0)}{\gamma LE}$$
(9.39)

where e_{sc} refers to the saturation vapour pressure at the surface temperature of the canopy and e_0 to the actual vapour pressure at this surface. Since it is rather difficult to define the surface of a canopy precisely, indirect methods are used. In neutral conditions when the shapes of the profiles are similar, air temperatures and vapour pressures are plotted against the wind velocity u at a number of heights within the boundary layer; the surface temperature and surface vapour pressure then correspond to the values at $u = 0$ (extrapolated from the velocity profile).

The *residual method*, also applicable to short term measurements, is based on eqns (9.35) and (9.38). Assuming that $r_{am} = r_{aw} = r_a$ then for a crop:

$$r_a + r_c = \frac{\rho c_p (e_{sc} - e_z)}{\gamma LE}, \quad \text{and} \quad r_a = \frac{\left[2 \cdot 5 \ln \left(\frac{z - d}{z_0} \right) \right]^2}{u_z}; \quad (9.40)$$

r_c is then determined by difference.

A rather simpler method, suited to longer term averages, is based on the ratio between the so-called potential evaporation E_p and the actual evaporation E. Potential evaporation here is defined as the evaporation when r_c is zero and is governed therefore, only by r_a. It may be measured when the canopy is wetted by rain. If r_a is known, then according to Monteith (1965):

$$r_c = r_a \left(1 + \frac{\Delta}{\gamma} \right) \left(\frac{E_p}{E} - 1 \right).$$
(9.41)

Alternatively, r_c may be calculated from the ratio of the loss of latent heat LE to the available net radiant energy $(R_N - G)$ according to an empirical relationship established by Monteith (1965) from measurements on a number of crops:

$$\log_{10} r_c = 1 \cdot 40 - \frac{2LE}{(R_N - G)}.$$
(9.42)

It should be appreciated that many of the above methods for determining r_c rest on the assumption that the profiles of latent heat (i.e. evaporation), sensible heat (i.e. temperature) and momentum (i.e. velocity) are identical, in other words that they originate at the same levels in the canopy. As the fluxes of latent and sensible heat are governed by foliar absorption of radiant energy, then provided that the stomatal resistances of the leaves in that part of the canopy do not change much, the heat and vapour sources will be similar, though seldom identical. For a crop with little foliage, when evaporation from the soil below makes an appreciable contribution to the total flux of water vapour from the area, anomalous values for r_c will result.

In the case of momentum, it is most likely that the sink (i.e. zero velocity) occurs at a higher level in the canopy than the sources of heat and water vapour suggesting that in reality, values for r_c should be corrected for this. A fuller analysis of the problem is given by Monteith (1973).

Relative importance of r_a and r_c

In the previous chapter, the relative importance of the leaf resistance, r_s and the air resistance, r_a was discussed in relation to the control of evaporation rates. We can now extend this discussion to crops in the field.

The values for r_c and r_a given in Table 9.4 serve to illustrate some of the more important relationships. Since r_c represents the combined resistances of all stomata in the canopy, acting more or less in parallel with each other it would be expected that it will decrease during the growing season as the leaf area per unit area of ground increases (the so-called *leaf area index*). This is clearly shown in the data for potato and lucerne. Diurnal increases in r_c, such as indicated by barley, reflect increasing stomatal closure as a result of increasing water stress as the day progresses.

Referring back to eqn (9.37) we see that for a given radiation income and vapour pressure deficit, evaporation will be largely determined by r_c when this is much larger than r_a, as for example when the stomata begin to close, and/or the wind speed is high. The relatively high r_c for the pine forest combined with a low r_a suggests that these trees exert a much greater stomatal control on transpiration than the other crops. When $r_c = r_a$, it seems that water loss is entirely controlled by r_a and hence by wind; this appears to be the general situation with the lucerne.

We can extend this argument to a generalization that when there is no water shortage and the stomata are fully open (i.e. when r_c is a minimum), transpiration from crops is almost entirely determined by weather conditions—the available net radiation, the vapour pressure deficit in the air and

TABLE 9.4
Crop and air diffusion resistances

		Apr.	May	June	July	Aug.	Sept.
		\multicolumn{6}{c}{Mean values for r_c and r_a in s cm^{-1}.}					
Mean wind speed (m s^{-1})		2·5	2·32	1·88	1·74	1·66	1·66
Pine forest	r_c	1·14	1·06	0·98	1·25	1·50	1·20
	r_a	0·03	0·03	0·03	0·03	0·03	0·03
Potato	r_c	1·30	1·26	1·13	0·45	0·60	1·10
	r_a	0·65	0·63	0·61	0·45	0·33	0·38
Lucerne	r_c	0·70	0·56	0·64	0·35	0·30	0·60
	r_a	0·54	0·44	0·60	0·43	0·33	0.52
		\multicolumn{6}{c}{Time of day}					
		6–7	8–9	10–11	12–13	14–15	16–17
*Barley	r_c	0·17	0·31	0·50	0·47	0·39	0·43

* From Szeicz and Long (1969); remainder from Szeicz *et al.* (1969).

the wind velocity. As we shall see this forms the basis of Penman's evaporation formula.

Evaporation formulae

An essential feature of the water regime of an area is the balance between the water income (rain, snow etc) and the amount lost by transpiration from vegetation, from evaporation from land and water surfaces, and where relevant, by drainage through the soil. A knowledge of the water balance of a catchment area is of considerable importance to water engineers in so far as it determines the flow of streams and the amount available for reservoir storage. It is also of vital importance to those concerned with the growth of crops since water deficiency is probably the most common factor limiting production.

Although data on the income of water at least as rainfall, is readily available in most areas, figures for evaporation are not as yet so common. Nevertheless, the importance of evaporation data has long been appreciated and numerous formulae have been proposed which attempt to provide such figures. Most of the formulae are so empirical that they are usually of little value except perhaps in restricted areas. Of the few that have proved to be of wider application, that of Penaman (1948, 1956) is one of the most successful. Most of the arguments underlying its derivation have already been covered.

Penman's formula is based on combined aerodynamic and energy balance principles and in its more recent form, is essentially the same as eqn (9.37) except that instead of the diffusion resistances, it incorporates factors that can more readily be obtained from standard meteorological data. With a slightly different notation than the standard form it can be expressed as:

$$E = \frac{(\Delta/\gamma)(R_N - G) + E_a}{\Delta/\gamma + 1}, \tag{9.43}$$

where as before, $(R_N - G)$ represents the net radiation energy available, Δ, the slope of the saturation vapour pressure–temperature curve at the mean air temperature (Fig. 9.10) and γ the psychrometric constant. The aerodynamic term E_a is a measure of the 'drying power' of the air and incorporates the effect of wind and vapour pressure deficit.

For convenience in its practical application, e.g. in the prediction of irrigation requirements, mean daily values are used and all energy units are expressed in terms of their evaporation equivalents of mm water (1 mm evaporation per day $= 2\cdot47 \times 10^6$ J m^{-2} per day).

Over the period concerned, G is assumed to be zero and R_N is calculated from:

$$R_N = E_s(1-\rho) - E_L,$$

where E_s is the income of solar radiation, ρ the reflection coefficient of the surface ($0\cdot25$ for a green crop such as grass) and E_L the balance of LW radiation exchange between the surface and the air. E_s is determined empirically from:

$$E_s = E_{sA}\left(a + b\,\frac{n}{N}\right)$$

where E_{sA} is the energy income at the top of the atmosphere for the given latitude and time of year (published data) and n/N the ratio of hours of bright sunshine to total daylight hours; the latter allows for the effect of cloud and is a standard meteorological measurement. The constants a and b vary with latitude; between 54°N and 56°N, commonly used values are $a = 0.155$ and $b = 0.69$.

E_L is calculated from the empirical relation:

$$E_L = \sigma T^4(0.47 - 0.075\sqrt{e})\left(0.17 + 0.83\,\frac{n}{N}\right) \text{ mm d}^{-1}$$

where σ is the Stefan–Boltzman constant and e the mean vapour pressure of the air (mm Hg) The quantities containing e and n/N allow for the effect of water vapour and cloud on radiation from the sky.

The aerodynamic term E_a is calculated from:

$$E_a = 0.35(1 + u/100)(e_s - e) \text{ mm d}^{-1}$$

where u is the daily mean run of the wind (miles per hour) at a height of 2 m above ground, and e_s the saturation vapour pressure at the mean air temperature (mm Hg).

The great practical advantages of this formula is that besides data from tables, only four weather attributes are required, the mean temperature and vapour pressure of the air, daily hours of bright sunshine and wind velocity. Specimen calculations with tables are published in the Ministry of Agriculture, Fisheries and Food Technical Bulletin No. 16 (1967). The British Meteorological Office regularly produces estimates of E based on a computer programme using slightly different constants than those given above.

It must be emphasised that the calculated evaporation rate E corresponds to *potential evaporation* or transpiration, i.e. water loss from a crop when it is plentifully supplied with water and there is no stomatal control. Under these circumstances, transpiration is governed almost entirely by external weather factors. Strictly speaking, the formula only applies to short grass covers for which the various empirical constants were originally derived; however, it is generally considered to apply also to most agricultural crops and even to deciduous trees, but not to conifers.

Drag forces

The drag exerted on air movement by a crop is accompanied by an equal reactive drag of the wind on the crop. A knowledge of the factors determining the magnitude of this force is of considerable practical importance because of damage to crops by wind. Lodging of cereals occurs when the stems are unable to withstand the combined forces of wind drag and the weight of the head (especially if wet) when the stem bends over. Shallow rooted trees are particularly susceptible to windthrow during gales.

From wind tunnel data, Tani (1963) concluded that rice stems would break when the bending moment about the base of the stems exceeded $0.2 \, \text{N m}^{-1}$ (moment $=$ force $\times$ distance from turning point). In the field however, only a moment of about $0.056 \, \text{N m}^{-1}$ was required; besides the possibility of disease affecting the strength of the stems, this could have been due to wind gusts with peak values much greater than mean values.

Fraser (1963) measured the drag on approximately 8m tall specimens of four conifer species in a large wind tunnel at air speeds ranging from 9 to $26 \, \text{m s}^{-1}$. At the lowest wind speed the relationship between the drag force and windspeed was curvilinear, but above this, the crowns deformed and became more and more streamlined; the drag force could then be expressed as a linear function of windspeed. Independent of windspeed, the drag increased linearly with tree weight and he was able to establish a significant regression of drag on windspeed, tree weight, and an interaction between these variables. To test the validity of these wind tunnel measurements in the field, estimates were made of the mechanical force required to uproot an 18 m tall Douglas fir; this required a moment of $8 \times 10^4 \, \text{N m}^{-1}$ about the base of the stem, corresponding to a wind tunnel speed of $25 \, \text{m s}^{-1}$. Similar trees were later uprooted by a gale with the same mean gust speed.

In later work (Walshe and Fraser 1963) model trees were used in a wind tunnel to determine how wind drag in plantations was affected by spacement and orientation.

The magnitude of the drag force is given by the surface shear stress τ_0, which is defined as a force per unit area of horizontal surface. In certain cases we might think of this as a skin friction force, but with vegetation and relatively high wind speeds, probably most of the drag arises as pressure drag on individual roughness elements such as leaves and stems.

According to our earlier definition, $\tau_0 = \rho u_*^2$ where u_* is the friction velocity, but since in the fully developed turbulent boundary layer, τ and u_* are assumed to be constant, we can apply this definition to the shear stress τ at any other convenient level within the turbulent zone, i.e. $\tau = \rho u_*^2$. It follows from eqns (9.14 and 9.15) that for a given windspeed u_z measured at a height z above the ground, the rougher the surface (as indicated by z_0) the greater will be the shear stress. Furthermore, given the values of u_z, z and z_0 (also d, where relevant), we can calculate u_* from these equations and then from $\tau = \rho u_*^2$, we can evaluate the drag force τ ($= \tau_0$).

Examples of these calculations are given in Table 9.3, first for the short grass and tall crop data illustrated in Fig. 9.6 and then for some data for a taller grass crop at different windspeeds. We see that with the same windspeed conditions, the drag on the tall crop is about 10 times that on the short grass crop. In the case of the tall grass crop, if the drag was mainly pressure drag, we would expect it to increase approximately as the square of the windspeed; that it has not quite done so must be attributed to bending over of the stems in the wind, as indicated by the progressive fall in z_0. By enclosing small areas of grassland in floating pans, Pasquill (1950) was able to measure directly the drag on the crop induced by wind and his results fully confirm the validity of the calculations based on eqn (9.15).

Roughness and turbulent transfer

Increasing roughness of a surface not only means a greater drag force, but also increased turbulent transfer since as explained earlier (p. 148), τ can be interpreted both as a shear stress per unit area and as a rate of momentum transfer.

From eqn (9.20a), putting $\tau = \rho u_*^2$, the turbulent transfer coefficient for momentum K_m is given by: $K_m = (u_*)^2/(\mathrm{d}u/\mathrm{d}z)$. Differentiation of eqn (9.15) gives

$$\mathrm{d}u/\mathrm{d}z = 2 \cdot 5 u_* / z - d$$

whence,

$$K_m = \frac{u_*(z-d)}{2 \cdot 5} \quad or, \quad K_m = \frac{u_* z}{2 \cdot 5} \quad \text{if } d = 0 \qquad (9.44)$$

(Note that K_m, though independent of the windspeed, varies with height at which the windspeed is measured.)

Using eqn (9.44), values of K_m at the given heights have been calculated from the data of Table 9.3. We see that at 4 m, K_m for the tall crop is over twice that for the short grass cover and that in both cases, it is of the order of 10^4 times that of ν the corresponding coefficient for the molecular diffusion of momentum

$$(\nu \text{ at } 20\,°\text{C} = 1 \cdot 5 \times 10^{-5} \text{ m}^2 \text{ s}^{-1}).$$

For the taller grass crop, K_m at 2 m height increases almost linearly with windspeed. Other factors being equal therefore, we would expect a more or less corresponding increase in heat and mass transfer by turbulent diffusion.

An alternative approach is to express the shear stress τ in terms of a drag coefficient C_D defined by $\tau = C_D \frac{1}{2} \rho u^2$ where u refers to the mean wind velocity at some convenient height, usually standardized at 2 or 10 m above the ground surface, (n.b. some texts, e.g. McIntosh and Thom (1969) define $\tau = C_D \rho u^2$) Since $\tau = \rho u_*^2$, then

$$C_D = 2 \left(\frac{u_*}{u} \right)^2 \quad \text{and} \quad \frac{u}{u_*} = \sqrt{\left(\frac{2}{C_D} \right)}. \qquad (9.45)$$

Substituting this in eqns (9.14 or 9.15) gives:

$$\sqrt{\left(\frac{2}{C_D} \right)} = 2 \cdot 5 \ln \left(\frac{z}{z_0} \right) \quad or, \quad \sqrt{\left(\frac{2}{C_D} \right)} = 2 \cdot 5 \ln \left(\frac{z-d}{z_0} \right). \qquad (9.46)$$

It follows that if C_D is referred to a particular height z and where relevant, d is constant, C_D is dependent only on the roughness length z_0, i.e. it provides an alternative measure of the roughness. When z_0 is constant C_D is constant and according to eqn (9.45), the ratio of the friction velocity (u_*) to the mean wind velocity (u) at the reference height is also constant. This allows us to determine C_D and therefore the drag at any given windspeed measured at the same height.

On the basis of eqn (9.45), values for C_D have been calculated from u_* and u in Table 9.3. That C_D for the tall crop is so much higher than that for the short grass cover confirms the much greater roughness of the former. Since z_0 for the tall grass crop is not constant, neither is C_D, but it varies much less than the other parameters.

As C_D is related to τ (by $\tau = C_D \frac{1}{2} \rho u^2$), then for a given windspeed it also provides a measure of the rate of momentum transfer, i.e. turbulent diffusion.

Sutton (1953) quotes data for C_D for various natural surfaces based on a windspeed of 0·5 m s^{-1} at 2 m height. These range from 0·002 for very smooth surfaces (mud flats, ice) to 0·032 for thick grass, 0·5 m tall. It is of interest that these values are of the same order of magnitude as the drag coefficients for large streamline surfaces with (Re) between 10^4 and 10^6 (Fig. 5.5). Drag coefficients have also been calculated for the trees mentioned earlier (Fraser 1963) and according to species, range from 0·32 to 0·72 at 9 m s^{-1}, falling to 0·16 to 0·4 at 26 m s^{-1} as the crowns become more streamlined (see also Mayhead, 1973). These cannot be compared with the drag coefficients for land surfaces because when dealing with individual trees, C_D is calculated from, *drag force* $= C_D \frac{1}{2} \rho u^2 A$ where A is the frontal area of the crown.

Windbreaks and shelterbelts

The function of windbreaks and shelterbelts is to create a 'sheltered' zone characterized by a reduced windspeed. In the case of crops this means a reduction in transpiration and a generally more favourable water balance, though concomitant effects on heat transfer may also contribute to better growth conditions. Because of their widespread use, shelterbelts have attracted a considerable literature, most of which is concerned with empirical observations on changes in the microclimate and crop responses. The following account is concerned primarily with the interpretation of shelterbelt effects in the light of the principles discussed in this and earlier chapters; comprehensive accounts including practical details are given in several reviews and texts e.g. Caborn (1957) and WMO (1964).

When wind encounters an impermeable barrier, its horizontal velocity is checked and the resulting increase in air pressure (Bernoulli effect) forces the oncoming air streams over the barrier (Fig. 9.13). As they do so, they are accelerated, the air pressure falls and under the pressure of the undisturbed air streams above, they are forced down again to regain their initial velocity

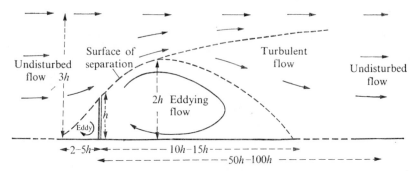

FIG. 9.13. Characteristics of the airflow pattern due to a near-impermeable cross-wind barrier. (Not to scale). From Gloyne (1954).

and direction at a distance of some 40 to 50 times the height of the barrier (n.b. since experiments have shown the extent of the shelter effect to be proportional to the height of the barrier, all distances are customarily expressed as multiples of barrier height). Leaving aside the small zone of disturbed air flow on the windward side, flow on the lee side may conveniently be divided into two regions: (*a*) a region of relatively large eddies induced by boundary layer separation at the top of the barrier and extending for about 10 to 15 times the barrier height; and (*b*) a rather more extended region in which the turbulence generated by the barrier gradually declines and the initial velocity is regained as a result of momentum transfer from the faster moving air streams above. The high turbulence behind the barrier detracts considerably from the shelter provided.

The important feature of shelterbelts used in practice is that they are partially permeable to the wind; air passing through the belt is slowed down, its pressure on the lee side increases, and as a result, turbulence is reduced and the recovery of the initial velocity delayed. In effect this means that the zone of reduced windspeed behind a permeable belt is more extensive than that behind an impermeable barrier. The effect of permeability (defined as the open area perpendicular to the wind, expressed as a percentage of the total belt area) on the distribution of the wind velocity measured at a fixed height is illustrated in Fig. 9.14.

We see that in all cases the wind-speed falls to a minimum at a certain distance behind the belt, varying according to the permeability, and then increases to attain its initial speed at a distance corresponding to about 40 to 50 times the height of the belt. This pattern is typical of the general behaviour of wind in the lee of shelterbelts, but the precise relationships vary according to the height at which the velocity is measured, the initial wind velocity and certain other factors (surface roughness, air stability) many of which are interrelated. Insufficient recognition of these factors or lack of the relevant quantitative information, has meant that many of the published field observations are of little more than local interest.

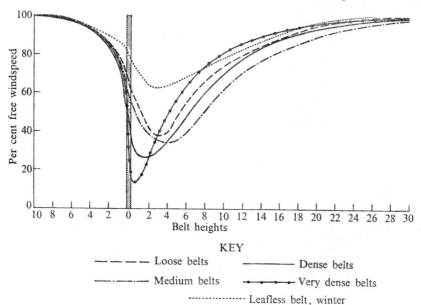

KEY

— — — — Loose belts   ———————— Dense belts

—·——·—·— Medium belts   •—•—•—•— Very dense belts

···················· Leafless belt, winter

FIG. 9.14. Effect of shelterbelt permeability on the pattern of relative windspeeds 1·4 m above the ground. From Nageli (1946).

As might be expected, the effect of permeability varies with the windspeed. With dense belts, increasing the wind speed extends the sheltered zone, but as the permeability increases, this is offset by a general increase in windspeed (note the behaviour of the leafless belt in winter). For a given wind velocity therefore, there should be an optimum permeability at which maximum shelter is achieved (i.e. the maximum distance over which the windspeed is reduced). For the conditions illustrated in Fig. 9.14 this corresponds to a belt of medium permeability.

The effect of permeability on the wind velocity profiles is illustrated in Fig. 9.15. With the denser belt, the velocity gradient within the height of the belt is relatively small so that the location of the minimum velocity does not change much with height. In the case of the more permeable belt, the velocity gradient is much steeper, and the nearer the ground, the further away is the minimum velocity located. With increasing roughness of the ground surface, we would also expect an earlier recovery of the wind velocity and a minimum nearer the belt. This has been confirmed by observations both in the field and in wind tunnel experiments.

The planning of shelterbelts requires a knowledge of the extent of the sheltered area, but as will be appreciated from the above account, this is difficult to define precisely because of the gradual recovery of the windspeed. A common approach to this problem is to define the effective sheltered zone as that in which the windspeed is reduced by 20 per cent. Depending on the permeability of the belt, Fig. 9.14 suggests a distance varying from about

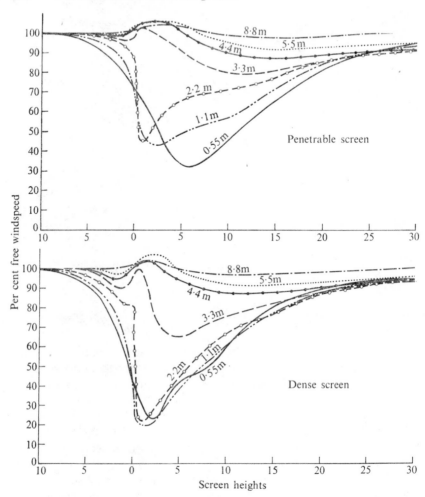

FIG. 9.15. Relative windspeeds at different heights above the ground in the vicinity of a permeable and a dense shelterbelt, 2·2 m high. From Nageli (1953).

15 to 20 times the height of the belt, but this distance may vary appreciably with wind velocity, the nature of the windbreak (i.e. the degree to which its permeability changes with windspeed) and the roughness of the surface. As far as the area of sheltered zone is concerned, account must also be taken of the length of the belt and the direction of the wind. According to Nageli (1953) the length over height ratio must be at least 11.5 if wind conditions are to approximate to those with an infinitely long belt; with a ratio of 2, the protected area is only about half the maximum area.

In view of the numerous variables affecting shelterbelt behaviour, the largely empirical approach to their design is not unexpected.

10. Non-Newtonian fluids

IN contrast to air and water, most biological fluids, including blood and lymph and 'semi-fluids' such as cytoplasm, are non-Newtonian; their viscosity is not independent of the applied shear stress and they therefore do not obey Newton's law of viscosity (eqn 1.1). The study of non-Newtonian fluids constitutes a major branch of the science of *rheology*.

Blood

We might begin with an experiment which illustrates the anomalous behaviour of blood compared with water.

If the volume rate of flow of water through a narrow glass tube is measured under different pressure heads, it will be found to obey Poiseuille's law (eqn 2.2) over the whole range, i.e. the flow rate will be directly proportional to the pressure head and the line relating these quantities will pass through the origin (Fig. 10.1). Since Poiseuille's law also states that the viscosity is inversely proportional to the ratio, flow rate/pressure head, and since the shear stress (τ) is directly proportional to the pressure head, the constant slope of this line confirms that the viscosity of water is independent of the shear stress and that water is a Newtonian fluid.

If the experiment is repeated with whole blood (defibrinated to avoid clotting) a linear relation will still be established between the flow rate and

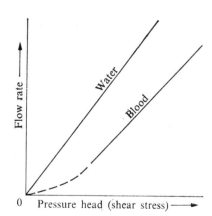

FIG. 10.1. Contrasting responses in the volume flow rate of water and blood in a capillary tube to increasing pressure head (shear stress).

pressure head (shear stress) at relatively high values, but the line does not extrapolate back to the origin. As shown in Fig. 10.1 when the shear stress falls below a certain value (in practice about 4×10^{-3} N m^{-2}, but varying with conditions), the relationship becomes curvilinear and the line appears to approach the shear stress axis, asymptotically. Over this region, as the shear stress is increased the slope increases and hence the viscosity decreases; within this range of shear stress therefore, blood behaves as a non-Newtonian fluid.

Composition and flow properties of blood

Blood consists essentially of a suspension of red cells (erythrocytes) in plasma. The proportion of other constituents such as white cells and platelets is usually so small that they are unlikely to affect the flow properties of blood under normal conditions. Though difficult to recognise *in vivo*, isolated red cells appear as biconcave discs, about 8 μm in diameter and 2 μm thick and in normal human blood they occupy about 45–50 per cent of the total volume (the so-called *haematocrit*).

Blood plasma consists of a solution mainly of three proteins, albumin, globulins, and fibrinogen, the last being responsible for clotting. There is some uncertainty as to whether fresh, untreated plasma alone is non-Newtonian (see Charm and Kurland 1974), but if blood is first allowed to coagulate and then centrifuged to remove the red cells, the remaining serum appears to behave as a Newtonian fluid. This, together with observations that normal red cells suspended in isotonic saline, still show non-Newtonian properties (at least in the capillary tube viscometer), suggests that the anomalous behaviour of blood must be largely, if not entirely, due to the suspended red cells.

The measurement of blood viscosity

Despite the considerable amount of experimental data on blood flow much uncertainty still remains concerning its interpretation. Part of the difficulty arises from the complicated nature of the problem, but confusion has also arisen because of probable errors in measurement. Up to the early 1960s, practically all measurements were made using the tube viscometer (p. 15) and defects in this technique have probably been the source of many misconceptions. For example, it has long been known that in glass tubes less than about 200 μm in radius, the apparent viscosity of blood, as determined from the pressure/flow ratio, decreases as the radius of the tube decreases. In a tube of radius 20 μm, corresponding to that of arterioles, the viscosity appears to be only about two-thirds of that measured in larger tubes such as are commonly used in viscometers. This effect, first described by Fahraeus and Lindquist in 1931 for blood and known as the *Fahraeus–Lindquist effect*, is the same as the *sigma* effect observed in other suspensions, e.g. latex and paint, in which the size of the particles is comparable with that of the tube radius. Vand (1948) explained the sigma effect on the argument that because no particle can be nearer to the wall than its radius, there must be a layer of medium (so-called slippage zone) next to the wall with fewer particles and hence with a lower viscosity. The smaller the tube the greater will be the relative effect of this slippage zone and lower the overall viscosity of the suspension.

The Fahraeus–Lindquist effect appears to be primarily due to a rather similar axial accumulation of the red cells, but although there is substantial evidence for this when blood flows in glass tubes, the situation *in vivo* is far less certain (see Charm and Kurland 1974). A narrow slippage zone probably occurs over a wide range of vessel diameters, but this is now generally believed not to be of any significance in explaining the anomalous behaviour of blood (Whitmore 1968). Cokelet (1972) has gone further and attributes the Fahraeus–Lindquist effect to a well known defect of the tube viscometer, namely that when blood flows steadily from a large reservoir into a small capillary, there is a 'skimming off' of the red cells with the result that the concentration of the red cells decreases as the capillary diameter decreases, thus causing a reduced viscosity.

Because of this and other defects of the tube viscometer most modern measurements with blood (and other non-Newtonian fluids) are made either with the coaxial or cone plate types of viscometer. As we shall see, this has led to expressions of fluid behaviour rather different to those adopted earlier, e.g. as in Fig. 10.1.

In the coaxial (or Couette) type of viscometer (Fig. 10.2) the fluid is held in a cylindrical pot which must be rotated at constant velocity. The viscous drag of the wall on the fluid is transmitted through the fluid to produce a torque (turning) force on the inner cylinder (the bob). This force can be measured either by the force required to return the bob to its original position, or if the bob is suspended on a wire, by its angular deflection.

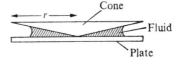

FIG. 10.3. The cone plate type of viscometer.

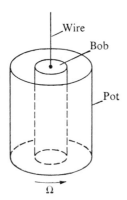

FIG. 10.2. The coaxial type of viscometer.

In effect this system corresponds to the movement of a fluid between two parallel plates and for Newtonian fluids with laminar flow, the torque T is given by:

$$T = \frac{4\pi r_1^2 r_2^2 h \eta \omega}{r_2^2 - r_1^2},$$ (10.1)

where r_1, r_2 correspond respectively to the radii of inner and outer cylinders, h to the height of the inner cylinder and ω to the angular velocity of the pot (in rad s^{-1}). For a given instrument and angular velocity therefore, the torque force T is proportional to the dynamic viscosity of the fluid η. If the instrument is calibrated in terms of these two quantities, the viscosity of an unknown Newtonian fluid can be determined directly from the torque force it produces at the same angular velocity.

Several precautions have to be taken when the instrument is used for blood or other non-Newtonian fluids and the dimensions and angular velocity should be such as to ensure that all the fluid is sheared. Since over a certain range at least, the viscosity of blood is no longer independent of the shear stress, eqn (10.1) cannot strictly be used. The simplest procedure is to assume that the fluid is Newtonian and to determine its apparent viscosity η_a at different shear rates from the relation:

$$\bar{\tau} = \bar{D} \eta_a,$$ (10.2)

where $\bar{\tau}$ and $\bar{D}$ are respectively the arithmetic mean shear stress and shear rate across the fluid. The arithmetic mean shear stress is given by:

$$\bar{\tau} = T\left(\frac{r_1^2 + r_2^2}{4\pi h r_1^2 r_2^2}\right).$$ (10.3)

Substituting for T from eqn (10.1) gives:

$$\bar{\tau} = \eta_a \omega \left(\frac{r_1^2 + r_2^2}{r_2^2 - r_1^2}\right),$$

whence from eqn (10.2),

$$\bar{D} = \omega \left(\frac{r_1^2 + r_2^2}{r_2^2 - r_1^2}\right).$$ (10.4)

In the cone–plate viscometer (Fig. 10.3) the fluid is contained in the space between a cone of very large angle (about 178°) and a flat plate at right angles to the cone; very little fluid is therefore required. Either cone or plate is rotated and the drag on one is recorded as a torque force by the other.

If α = angle between plate and cone (radians), r = radius of cone and plate, ω = angular velocity of rotation (radians per sec) and T = torque force, then:

$$\text{rate of shear } D = \frac{\omega}{\alpha}$$

and

$$\text{shear stress } \tau = \frac{3T}{2\pi r^2}.$$

Hence for a Newtonian fluid, the viscosity

$$\eta = \frac{\tau}{D} = \frac{3T\alpha}{2\pi r^2 \omega}. \tag{10.5}$$

The linear relationship between the torque force T and the viscosity η allows the instrument to be calibrated directly in terms of the viscosity. In the case of non-Newtonian fluids, eqn (10.5) gives an apparent viscosity η_a.

Because of the way in which the measurements are made with the coaxial and cone plate types of viscometer, it is customary to express the behaviour of blood and other non-Newtonian fluids in terms of either the relationship between the apparent viscosity and the shear rate (D) or the relationship between the shear stress (τ) and the shear rate. There appears to be no special advantage of these methods of expression other than practical convenience. Typical results for the apparent viscosity of blood as a function of the shear rate are illustrated in Fig. 10.4 and clearly show the asymptotic decrease in η_a as the shear rate is increased.

The cause of non-Newtonian behaviour of blood

Considerable attention has been given to the origin of the non-Newtonian behaviour of blood; that the problem is still not yet entirely resolved is largely

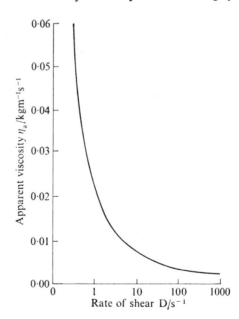

FIG. 10.4. Effect of rate of shear on the apparent viscosity of blood. Redrawn from data of Whitmore (1968), Fig. 6.4.

due to the fact that several factors appear to be involved, many of them interrelated, such as shear rate, size of vessel, rate of flow and the deformation, aggregation, and orientation of the red cells.

The anomalous behaviour of normal blood, as indicated in Fig. 10.4 only appears at low shear rates, below about 50 s^{-1}, and in the smaller vessels. Above about 100 s^{-1}, blood behaves almost like a Newtonian fluid and its apparent viscosity then remains fairly constant and independent of the shear rate at between 0·003 and 0·004 kg m^{-1} s^{-1}. That this is only about half the viscosity of a suspension of smooth rigid spheres at the same concentration is believed to be due to the flexibility of the red cells. This view is supported by the observation that up to a haematocrit of 40 per cent, suspensions of red cells, hardened by formaldehyde, are no longer non-Newtonian.

At very low shear rates, when the red cells tend to behave more like rigid bodies, the viscosity of blood is significantly higher than a suspension of oblate ellipsoids of the same concentration. One explanation is that the cells immobilize some of the plasma, thus increasing the particle concentration and therefore the viscosity, but other factors have to taken into account.

At these low shear rates the red cells form clumps which must be disrupted before flow can occur. The force required, the so-called *yield stress*, is probably small, but its existence would explain why the flow rate × shear stress curve shown in Fig. 10.1 does not extrapolate back to zero; the intercept on the shear stress axis would then correspond to the yield stress. As the shear stress is increased, the aggregates are broken up into smaller and smaller groups and eventually to single cells; along with the release of hitherto immobilized plasma, this results in a decreased concentration of red cells and hence in a lower viscosity. At this stage blood behaves as a Newtonian fluid.

The most common form of aggregation at low flow rates is the *rouleau* in which a number of cells adhere face to face, but there may also be a random aggregation with no obvious structure. *In vivo* aggregation is most clearly seen in the venules where the cells emerge from a number of capillaries in a rather irregular fashion. In the larger veins and arteries the flow rates are probably fast enough and the shear rates high enough to ensure that aggregation no longer makes a significant contribution to the viscosity.

At relatively higher shear stresses (above about 0·5 N m^{-2}) the red cell membrane appears to rotate around the cell contents; the cell then assumes the properties of a fluid drop and whole blood behaves more like an emulsion than a particulate suspension. This has been suggested as the way in which red cells adapt themselves to flow (Charm and Kurland 1974).

As the vessel size decreases, the red cells arrange themselves more and more in single file and when it approaches that of the red cells, as in the capillaries (in man about 8 μm in diameter), they then move axially with their faces perpendicular to the direction of flow, either in short trains (*plug flow* or *axial train flow*) or individually (*bolus flow*). Because of their flexibility the cells

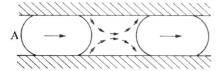 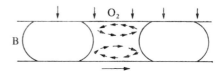

FIG. 10.5. Bolus flow. (A) Depicts the actual motion of the plasma between the red cells. The speed of the plasma at the axis is about twice that of the red cells; when it reaches the red cell in front it is diverted towards the wall where it remains until the arrival of the red cell behind. Relative to the red cells, the motion of the plasma is circular as shown in (B). This mixing motion greatly facilitates equilibration with gases diffusing through the wall. From Physiology and Biophysics of the Circulation, 2nd edition, by Burton, A. C. Copyright © 1972 by Year Book Medical Publishers, Inc. Chicago. Used by permission.

can adjust their shape to the size of the smallest capillary leaving a thin layer of plasma next to the wall.

By injecting dye into a model bolus system of air and water, Prothero and Burton (1961; 1962) demonstrated appreciable eddy motion within the fluid (Fig. 10.5). Assuming that the same occurs in the plasma between the red cells for blood flow *in vivo*, they concluded that bolus flow would be twice as effective as Poiseuille flow for the exchange of heat between the cells and the walls and that gaseous exchange would also be accelerated.

The circulation of blood

The following account of the dynamics of blood circulation is given mainly to place the flow properties of blood into perspective. The mean pressure of the blood leaving the left ventricle of the heart in man is about $1 \cdot 33 \times 10^4$ N m^{-2} (100 mm Hg), but by the time it returns to the heart its pressure has fallen to about one tenth of this. The greatest resistance to flow, and hence the greatest pressure drop occurs in the micro-circulation, especially in the arterioles and capillaries. The velocity of flow ranges from about $0 \cdot 1$ m s^{-1} in the arteries (1 to 10 mm in diameter) to less than 10^{-3} m s^{-1} in the capillaries (about 8 μm in diameter). Assuming for whole blood a dynamic viscosity of $0 \cdot 0035$ kg m^{-1} s^{-1} and an average density of $1 \cdot 06 \times 10^3$ kg m^{-3}, the Reynolds' number (based on the diameter and assuming a long straight tube far from an entry) ranges from about 4000 in the ascending aorta (systolic velocity $0 \cdot 63$ m s^{-1}, diameter $0 \cdot 02$ m) to between 900 and 150 for the larger arteries and veins and down to the order of $0 \cdot 001$ to $0 \cdot 003$ for the capillaries (n.b. some physiologists still base the Reynolds' number on the tube radius so that their values would be half the above). We might therefore expect turbulent flow in the aorta, but only laminar flow in the smaller vessels. However, according to Charm and Kurland (1974), the critical (Re) for blood flow in straight tubes, 2 mm or less in diameter, is only about 800, so turbulence might be expected in arteries larger than about 3 mm in diameter, but not in those less than 1 mm. Furthermore, as has already been mentioned

(p. 19) the occurrence of branches induces turbulence and laminar flow in pre-branch vessels invariably becomes turbulent after passing a junction.

Apart from the extra resistance to flow associated with turbulence, it seems that the higher transfer of energy from the blood stream to the wall and the resulting vibrations in the latter (*murmur*) may be symptomatic of vascular pathology (Burton 1972).

The outstanding feature of arterial flow is its pulsatory nature. Because the walls of the arteries are thick and elastic (to withstand the pressure) they will expand and contract, as a result of which a pressure wave is propagated at a rate some 10 to 20 times faster than the rate of blood flow. However, the gain in velocity during diastolic flow, when flow is augmented by the elasticity of the walls, may be more than offset during systolic flow when some energy is expended in distending the walls; the net result is that the mean flow rate is about the same, or perhaps slightly lower than if flow was steady.

The situation is very different in the smaller vessels and capillaries. By the time the blood has reached these, pulsatory flow has been largely damped out and in the absence of a pressure wave, the walls need only be thin; this of course enhances the exchange of heat and gases through the walls. At the low Reynolds' numbers prevailing, viscous forces predominate and the small size of the vessels makes the particulate nature of blood much more important.

Energy relations

The mechanical work done by the heart in pumping blood around the circulatory system is determined by the product of the pressure and the volume flow rate. In the case of the ventricle ejecting blood into the artery, both the pressure and volume (the *stroke volume*) vary and a complete record of these changes is necessary before the work can be calculated. We shall not go into the various short cuts adopted by physiologists, but will only consider the data quoted by Burton (1972) according to which each ventricle has an output of 8.3×10^{-5} m^3 s^{-1} and the mean pressure in the left ventricle is 1.33×10^4 N m^{-2}; the mechanical power output of the left heart is then 1.1 W. The mean pressure in the right ventricle is only about $\frac{1}{5}$th of that in the left, so its power output will be about 0.2 W, giving a total heart output of 1.3 W. During heavy exercise the volume output may increase by about 4 times the above and the mean pressure by about 50 per cent, giving a maximum power output of 7 to 8 W. This is a surprisingly low value, but its significance only becomes apparent when we consider the mechanical efficiency of the heart, i.e. the ratio of the mechanical work done to the total energy expended (as estimated from the rate of O$_2$ consumption). The latter includes energy expended in muscular activity and it turns out that for the heart, something like 90 per cent or more of the energy is dissipated in this way.

At rest, the mechanical efficiency of the heart is very low, usually of the order of 3 per cent, but it increases to a maximum of 10 to 15 per cent as the work load increases. This increase in efficiency follows from the ratio defining the mechanical efficiency: doubling the mechanical work load will double the numerator, but has relatively little effect on the denominator since the contribution of the energy utilized for mechanical work is so small. This knowledge of heart energetics is extremely important in the treatment of cardiac patients. Since the total load on the heart muscle (i.e. O_2 consumption) is the important factor, not the mechanical load, it is essential to control the arterial blood pressure (*hypertension*) and any factor that increases the duration of heart muscle contraction rather than to avoid mild exercise of the dynamic type (see Burton 1972).

Other non-Newtonian fluids

The flow relations of most non-Newtonian fluids can be expressed empirically by a power law of the general form:

$$\text{shear stress } (\tau) = k(\text{shear rate})^n \tag{10.6}$$

where k is a proportionality factor and n may vary between 0 and $1+$.

This equation has no real physical significance, but is useful for distinguishing between different categories of non-Newtonian fluids and for expressing experimental data. (Fig. 10.6).

When $n = 1$, eqn (10.6) reduces to eqn (1.1) for a Newtonian fluid with $k =$ viscosity η (Fig. 10.6, curve A).

Bingham-plastic fluids

These (Fig. 10.6, curve B) include plastic muds and suspensions of regular granular solids. When at rest they form a fairly stable 3-dimensional structure which can only be deformed by applying a certain shear stress; below this they behave as solids, above, as Newtonian liquids.

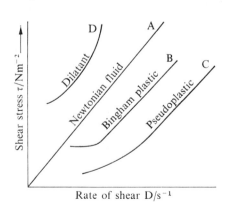

FIG. 10.6. Categories of non-Newtonian fluids, with a Newtonian fluid for comparison.

Pseudo-plastic substances

These are shown in Fig. 10.6, curve C with n between 0 and 1; they include solutions and suspensions of large elongated molecules and asymmetric particles, systems which characterize most biological fluids (e.g. blood, milk), also colloids such as clay and gelatin. Over a certain range at least, their viscosity decreases with increasing shear rate, in many cases to a much greater extent than blood; this behaviour is generally believed due to entanglement or aggregation of the particles at low shear rates, with separation and alignment at higher rates.

Dilatant fluids

As shown in Fig. 10.6, curve D, with n greater than 1, these increase in viscosity with increasing shear rate. This behaviour is typically shown by an aqueous suspension of finely-ground quartz (1–5 μm particle diameter, 44 per cent solid); a glass rod can easily be moved through the suspension at low speeds, but encounters an enormous resistance when the speed is increased beyond a certain limit. The effect of dilatancy is also seen when walking over wet sand. Similar behaviour is shown by pastes of intact starch grains in water, by concentrated sugar solutions and perhaps most strikingly, by 'bouncing putty'. A common feature of dilatant fluids is that at a low shear stress (low velocity) they behave like Newtonian fluids but more like a solid at higher values. The explanation commonly advanced is that at a low shear stress, there is just enough liquid to fill the voids between the particles; when a high stress is applied, the particles move over each other and occupy more space, thus forcing liquid out (dilatancy). Since there is no longer enough liquid to fill the voids, the apparent viscosity increases.

With several non-Newtonian fluids, the viscosity for a given shear stress changes with the length of time the shear stress is applied. When the viscosity increases with time, the fluid is known as *rheopectic*, and when it decreases, as *thixotropic*. In other words the viscosity of thixotropic fluids decreases when, and for as long as a mechanical force or shear stress is applied. Most readers will be familiar with thixotropic emulsion paints which are gel like when at rest, but flow readily as long as they are brushed. Thixotropic behaviour is shown by aqueous gels of certain mineral oxides (Al, Fe, etc), but of much greater significance from the biological point of view, by colloidal suspensions of proteins, including cytoplasm, synovial fluid, and gelatin.

The rheology of cytoplasm

Because of its fundamental importance we might take a closer look at the properties of cytoplasm. From a gross rheological point of view, cytoplasm may be regarded as a pseudoplastic colloidal suspension of proteins in an aqueous medium, showing pronounced thixotropic behaviour. Much of the classical work on the rheology of cytoplasm was carried out before electron microscopy had revealed a much more complicated membranous structure (endoplasmic reticulum etc.) than had hitherto been suspected (Seifriz 1942; Frey-Wyssling 1952). Nevertheless, we can still usefully distinguish between the semi-liquid sol state classically identified with endoplasm and the semi-solid gel state, with ectoplasm.

The relatively low viscosity characterizing the sol state is clearly revealed in the phenomenon of *endoplasmic streaming* (cyclosis) such as is commonly observed in the cells of many organisms, e.g. the rhizoid cells of the alga, *Nitella*. In contrast, the ability of animal cells to maintain a well-defined form implies a fairly rigid gel-like structure in the ectoplasm. Observations on the movement of included iron filings in a magnetic field have confirmed the much higher viscosity of the gel state. The elasticity of ectoplasm may be demonstrated in several ways, e.g. if an iron filing is implanted into the outer cortical layer of certain marine eggs and then displaced by a magnet, it will return to its original position when the magnetic field is removed.

Marsland (1942) describes several experiments in which the application of a hydrostatic pressure has induced the conversion of cytoplasm from the gel to sol states, thus confirming its thixotropic behaviour. Certain processes which depend on the structural characteristics of the gel state, or on sol–gel transformations, such as amoeboid movement (Allen and Francis 1965), streaming, and cell division (in marine eggs) are particularly sensitive to pressure-induced inhibition, and the degree of inhibition appears to be proportional to the degree of solation, which in turn, is proportional to the applied pressure. Up to a pressure of about 270 bars, provided that it is not maintained for more than about 1 hour, the effect is reversible; when the cells are returned to atmospheric pressure, the original characteristics of the gel are restored, usually within a minute or so.

The ingenious experiments of Kamiya (1942) nicely illustrate the close relationship between streaming and pressure. If a small mass of the slime mould, *Plasmodium* is divided into two parts connected by a strand of cytoplasm (Fig. 10.7), endoplasmic streaming occurs in the strand, oscillating in rather a complex manner from one mass to the other. Kamiya was able to accelerate, retard or even reverse this streaming by applying an appropriate

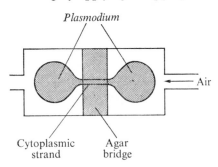

FIG. 10.7. Apparatus of Kamiya (1942) to investigate the effect of pressure on cytoplasmic streaming in the slime mould, *Plasmodium*.

air pressure difference to the two masses; to stop streaming, a pressure difference of up to about 2000 N m⁻² (20 cm water) was required. Hyashi (1961) gives a useful account of this and other experiments on cytoplasmic movement in cells.

The origin of the gel structure

The rigidity and elasticity of the cytoplasmic gel clearly indicates the presence of an internal structure. Cytoplasm contains about 10 per cent protein which, bearing in mind that ordinary table jelly can form a fairly rigid gel with only about 2 per cent protein, would appear to be adequate. However, the two substances cannot strictly be compared because of differences in the nature of the proteins involved. Table jelly is formed from gelatin by treatment with boiling water which denatures the proteins. Gelatin obtained from collagen typically contains mostly fibrous proteins which in their normal state adopt a coiled (helical) conformation (Fig. 10.8a). This structure is destroyed in denaturation, allowing the protein threads to intermingle and cross link so producing a 'brush heap' type of network which imparts rigidity to the system (Fig. 10.8b). Cytoplasm on the other hand is believed to contain mostly globular proteins which characteristically form ball-like structures (Fig. 10.8c). Precisely how these form a continuous network consistent with the properties of ectoplasm is still the subject of speculation. The various theories advanced include: (a) aggregation of the globular molecules into chains or into microtubules (Fig. 10.8d); the latter have been identified in the cytoplasm of certain cells at least and since they produce a fibre-like structure, they may at least contribute to the framework, and (b) linkage between the protein molecules by ordered water molecules which, as already explained, ensheath all macromolecules in an aqueous medium.

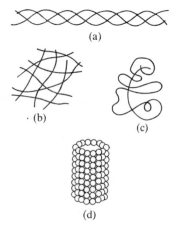

(a)

(b)

(c)

(d)

Fig. 10.8. Molecular structures. (a) 3-stranded molecule of collagen (b) 'brush heap' network of protein strands (c) folded chain of a globular protein (d) hypothetical model of a microtubule formed from globular proteins.

11. Some thermodynamic concepts of water movement

LIKE all other forms of matter, fluids obey the laws of thermodynamics. The fact that this has not been emphasized before is due partly to the difficulties that most biologists experience with thermodynamics and partly to the availability of other approaches. In the case of water movement however, plant physiologists in particular now make much use of thermodynamic arguments and terminology. It is therefore necessary to take some account of the principles involved, though no attempt will be made here to go into any details; these are more than adequately covered in the numerous texts on the subject.

Free energy and chemical potential

Since our main interest is in water movement we are primarily concerned with that part of the total internal energy of a system, known as *free energy*, that can be harnessed to do useful work. In the present context an example of such useful work would be osmotic work. If the underlying process is reversible, the free energy that can be obtained from a system at constant temperature and pressure is called *Gibbs free energy* (G) and the amount of Gibbs free energy per mole of each substance present in the system defines the *chemical potential* (μ) of that substance. In practice it is often more convenient to work in terms of the decrease in Gibbs free energy ($-\Delta G$) than in absolute terms since this provides a measure of the work that has been obtained from the system. Per mole of substance therefore, $-\Delta G$ will correspond to a change $-\Delta\mu$.

For an ideal gas, the chemical potential at a given absolute temperature T is determined by its pressure p:

$$\mu = RT \ln p - \mu^*, \tag{11.1}$$

where R = gas constant and μ^* refers to the chemical potential of the gas at the same temperature, but at unit pressure. It follows that the difference in chemical potential between a gas at pressure p_1 and the same gas at pressure p_2 (at the same temperature) will be:

$$\mu_1 - \mu_2 = RT[\ln p_1 - \ln p_2] = RT \ln\left(\frac{p_1}{p_2}\right). \tag{11.2}$$

Now suppose that instead of a pure gas we have a mixture of gases, e.g. air and water vapour. When there is no chemical interaction, each gaseous constituent of the mixture will behave independently and will contribute a certain amount of free energy (so-called *partial Gibbs free energy*) according to its partial pressure p. The total free energy of the mixture is then the sum of the partial free energies of the constituents.

We can now assign a chemical potential to each constituent on the basis of its free energy per mole (the *partial molal free energy*). If the chemical potential of water vapour in unsaturated air is μ_w and that in saturated air (at the same temperature and total air pressure) is μ_w°, the difference will be as in eqn (11.2):

$$\mu_w - \mu_w^\circ = RT \ln\left(\frac{p}{p_s}\right) \tag{11.3}$$

where p, p_s refer to the partial pressures of water vapour in unsaturated and saturated air respectively. Since these partial pressures define the respective vapour pressures e and e_s (the latter being the saturation vapour pressure at the absolute temperature T), and since the ratio e/e_s defines the relative humidity h we can rewrite eqn (11.3) as:

$$\mu_w - \mu_w^\circ = RT \ln\left(\frac{e}{e_s}\right) = RT \ln(h). \tag{11.4}$$

The chemical potential of a substance is an intensive property like temperature. Given the opportunity to do so, the substance will move spontaneously from a state of higher to lower potential, irrespective of the phase, until the potentials are equal. In other words, evaporation will occur from water until at equilibrium, the potential in the vapour phase equals that in the liquid phase. The potential of pure (liquid) water therefore, will be the same as that of water vapour in saturated air at the same temperature.

We know that the vapour pressure in equilibrium with a salt solution is lower than that over pure water; hence we might deduce that water in a solution has a lower chemical potential than that of pure water at the same temperature. If the equilibrium vapour pressure over a salt solution is e and that over pure water is e_s, then according to Raoult's law:

$$\frac{e_s - e}{e_s} = x_s \quad or, \quad \frac{e}{e_s} = 1 - x_s, \tag{11.5}$$

where x_s is the mole fraction of solute, i.e. the ratio of the number of moles of solute to the total number of moles present in the solution. x_s will of course increase as the solute concentration increases.

Replacing e/e_s by $1 - x_s$ in eqn (11.4) gives:

$$\mu_w - \mu_w^\circ = RT \ln(1 - x_s), \tag{11.6}$$

and since to a first approximation, $\ln(1 - x_s) = -x_s$ we can rewrite this equation as:

$$\mu_w - \mu_w^\circ = -RTx_s. \tag{11.7}$$

In other words at a constant temperature, the difference in the chemical potential of water in a salt solution and that in pure water increases as the salt concentration. Strictly speaking this only applies to dilute solutions, but if x_s is multiplied by an activity coefficient y, the product yx_s (the *activity*) validates eqn (11.7) for all concentrations. Data on y for various solutes at different concentrations are given in most texts on Physical Chemistry.

If the mole fraction of solute in aqueous solution is x_s and the only other constituent of the solution is water, the mole fraction of water will be $1 - x_s$. Putting $1 - x_s = x_w$ in eqn (11.6) allows us to define the difference in chemical potential solely in terms of the water fraction:

$$\mu_w - \mu_w^\circ = RT \ln x_w. \tag{11.8}$$

Water potential

In all the above equations μ is expressed in terms of energy per mole, i.e. N m mol^{-1}; dimensionally this is the same as a pressure (N m^{-2}) $\times$ the volume (m^3) per mole. In plant water relations it is much more convenient to work in units of pressure and this can be readily obtained by dividing μ by the volume occupied by 1 mole (the *molal volume*). In the case of mixtures or solutions, the molal volume of a constituent is known as the *partial molal volume*. ($\bar{V}$). Applying this to eqn (11.8) for water we have:

$$\frac{\mu_w - \mu_w^\circ}{\bar{V}} = \frac{RT}{\bar{V}} \ln x_w = \psi. \tag{11.9}$$

The new quantity ψ is called the *water potential* and as already explained, is expressed in pressure units (N m^{-2}). For most practical purposes, $\bar{V}$ for water (weight of one mole, 0·018 kg) $= 0·018 \times 10^{-3}$ m^3 ($= \bar{V}_w$).

By definition ψ for pure water is zero (since $x_w = 1$, $\ln x_w = 0$); hence the water potential of solutions at the same temperature and pressure (assuming no difference in altitude) will be negative and the higher the salt concentration, the more negative it will be.

We can apply exactly the same arguments to water vapour in air. If, as before, the vapour pressure of unsaturated air is e and that of saturated air at the same temperature is e_s, then dividing eqn (11.4) by $\bar{V}_w$ and comparing with eqn (11.9) we obtain an expression for the water potential of unsaturated air:

$$\psi = \frac{RT}{\bar{V}_w} \ln(h), \tag{11.10}$$

where again, $h = $ relative humidity.

The advantage of expressing potentials in pressure units becomes clear when we consider an osmotic system in which pure water ($\psi = 0$) is separated from say, a sugar solution by a semi-permeable membrane. Because ψ for pure water is higher than that for the solution, water will move spontaneously into the solution until in theory, the water potentials on both sides of the membrane are the same. In practice, when the solution is enclosed or connected to a manometer, movement will continue until it is prevented from doing so by the back hydrostatic pressure. We recognize this pressure as the osmotic pressure (π) of the solution. In effect therefore, equilibrium has been established by applying a pressure π to the solution such as to make up for the difference in water potential between pure water and the solution. If the latter is ψ, the difference is $0-\psi$, whence:

$$0-\psi = \pi \quad \text{or} \quad \pi = -\psi. \tag{11.11}$$

In other words, the water potential of a solution is the same as its osmotic pressure, but with a negative sign.

From eqn (11.7) we can define ψ and π for a solution in terms of its salt concentration x_s:

$$\psi = \frac{\mu_w - \overset{o}{\mu_w}}{\overline{V}_w} = -\frac{RT}{\overline{V}_w} x_s = -\pi. \tag{11.12}$$

For very dilute solutions, i.e. when x_s is very small, $x_s/\overline{V}_w \approx c_M$, the salt concentration expressed as moles per unit volume. Hence eqn (11.12) can be expressed in the more familiar form:

$$\pi = RTc_M. \tag{11.13}$$

Neither eqn (11.12) nor eqn (11.13) are valid for electrolytes unless account is taken of dissociation, besides the activity coefficients of the ions.

From the fact that equilibrium in an osmotic system can be established by applying a pressure to the solution, it follows that the water potential of a system is increased by an amount equivalent to the applied pressure. This contribution to the water potential is known as the *pressure potential*. Conversely, if the pressure on a system is reduced, i.e. it is subjected to a tension or suction, the potential is reduced accordingly.

Matric potential

When water or solution is present in a porous capillary system such as soil or a plant cell wall, we meet another factor contributing to the water potential known as a *matric potential*.

Imagine a capillary tube of radius r standing in a beaker of pure water and that equilibrium has been established when the water has risen to a height z above the water level outside. (Fig. 11.1). Assuming for the sake of simplicity that the contact angle is zero, then if the surface tension is σ (N m^{-1} or J m^{-2}), the force acting on the perimeter of the meniscus ($= 2\pi r\sigma$) must balance the weight of the water column ($= \pi r^2 \rho_w gz$ where $\rho_w =$ density of

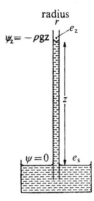

FIG. 11.1. Rise of liquid in a capillary tube. For details see text.

water and g = acceleration due to gravity). Hence:

$$2\pi r\sigma = \pi r^2 \rho_w g z \qquad or, \qquad z = \frac{2\sigma}{\rho_w g r}. \tag{11.14}$$

The capillary rise therefore, is inversely proportional to the radius of the capillary.

By definition, the potential of the water at the bottom of the capillary, coinciding with the water surface outside, is zero. The difference in hydrostatic pressure between this level and that at height z corresponding to the bottom of the meniscus = $\rho_w g z$; hence the water potential at this height (ψ_z) is given by:

$$\psi_z = -\rho_w g z = -\frac{2\sigma}{r} \qquad \text{(from eqn 11.14).} \tag{11.15}$$

At equilibrium ψ_z must also be the water potential of the air just above the meniscus and at the same level outside the capillary tube. If the relative humidity of the air at this level is h, then according to eqns (11.10 and 11.15):

$$\psi_z = \frac{RT}{\bar{V}_w} \ln(h) = -\rho_w g z. \tag{11.16}$$

With $\rho_w = 10^3$ kg m^{-3} and $g = 9\cdot81$ m s^{-2}, $\psi_z \approx 10^4$ N m^{-2} per m; in other words, in a column or water, the water potential falls by about 10^4 N m^{-2} (about $0\cdot1$ bar) for each m rise in height.

Since $\bar{V}_w$ = volume of 1 mol of water and ρ_w = density of water,

$$\rho_w \bar{V}_w = M_w, \qquad \text{the molecular weight of water.}$$

We can therefore write:

$$RT \ln(h) = -M_w g z,$$

and substituting for z from eqn (11.14) we have:

$$\ln(h) = -\frac{2\sigma M_w}{RTr\rho_w}. \tag{11.17}$$

Taking the following values at 20 °C ($T = 293$ K):
σ(water–air) $= 73 \times 10^{-3}$ N m^{-1}, $M_w = 18 \times 10^{-3}$ kg, $\rho_w = 10^3$ kg m^{-3},

$$g = 9 \cdot 81 \text{ m s}^{-2} \quad \text{and} \quad R = 8 \cdot 31 \text{ J mol}^{-1} \text{ K}^{-1},$$

we obtain for eqns (11.14, 11.15, and 11.17):

$$z = \frac{1 \cdot 49 \times 10^{-5}}{r} \text{ (m)}$$

$$\ln(h) = -\frac{1 \cdot 078 \times 10^{-9}}{r} \quad \text{or} \quad \log_{10}(h) = -\frac{4 \cdot 69 \times 10^{-10}}{r}$$

$$\psi_z = -\frac{0 \cdot 146}{r} \text{ (N m}^{-2}\text{)}.$$

TABLE 11.1

Capillary rise (z), equilibrium relative humidity (h) and water potential (ψ) for tubes of radius (r)

r m	z 1·49 m	h	$-\psi$ 1·46 Nm^{-2}
10^{-3}	10^{-2}	1·000	10^2
10^{-4}	10^{-1}	1·000	10^3
10^{-5}	1	1·000	10^4
10^{-6}	10	0·999	10^5
10^{-7}	10^2	0·989	10^6
10^{-8}	10^3	0·898	10^7
10^{-9}	10^4	0·340	10^8
10^{-10}	10^5	0·000021	10^9

Values for z, h, and ψ_z for different values of the capillary radius (r) are given in Table 11.1. The significant features of these data are: (1) r has to be below about 10^{-7} m (0·1 μm) before the equilibrium relative humidity of the air is appreciably affected, but beyond this, h falls very rapidly; (2) the very low potentials that can be developed in small capillary systems; in principle the same applies to any medium that imbibes water even when this is not strictly due to capillary action. These matric potentials, as we shall see later, play a very important role in plant and soil water relations.

At this stage it is useful to summarize the various factors that determine the water potential in systems of biological interest.

(1) *Solute or osmotic potential* ψ_s; equal to the negative osmotic pressure of the solution or medium, cf. eqn (11.11).

(2) *Pressure potential* ψ_p; equal to the difference in pressure between a standard pressure (usually atmospheric) and that prevailing in the system. A positive pressure increases the potential, a negative pressure or tension decreases it.

(3) *Hydrostatic or gravitational potential ψ_g*; this is related to the pressure potential, but refers specifically to pressures or tensions produced by a column of water and equals about 10^4 N m^{-2} per metre length of column.

(4) *Matric potential ψ_m*; the potential of water present in porous, capillary or otherwise imbibing media; in the former, decreasing inversely with the pore radius, cf. eqn (11.15).

In any given system more than one of the above components may be involved, in which case the total water potential is the arithmetic sum of these components. In a turgid vacuolated plant cell for example, the potential of the vacuolar sap will be the sum of an osmotic potential ψ_s ($= -\pi$, the osmotic pressure due to solutes) and a pressure potential ψ_p due to the pressure of the cell wall ($=$ turgor pressure). It follows that the true osmotic pressure of the vacuolar sap can only be measured when the turgor pressure is zero, i.e. at incipient plasmolysis; at this stage the water potential of a tissue will be equal in magnitude to its osmotic pressure.

It will be appreciated that the system as a whole will be in equilibrium with air at a relative humidity corresponding to:

$$\psi = \frac{RT}{\bar{V}_w}\ln(h).$$

This provides the basis of a method for determining the water potential of a tissue, namely by measuring the relative humidity of the air in equilibrium with the tissue (e.g. by a thermocouple psychrometer).

Water transport in the soil–plant–atmosphere system

The adoption of a thermodynamic approach based on water potentials has led to a far better understanding of the principles underlying the movement of water from the soil through a plant into the surrounding air. The following account is concerned mainly with these principles; further details can be obtained from several specialist texts, e.g. Slatyer (1967), Kramer (1969).

It is now generally believed that the major force responsible for the transport of water, in tall plants at least, derives from the evaporation of water from the walls of the mesophyll cells within the leaves. Cell walls are typically composed of a matrix of cellulose fibrils around which other substances are deposited (pectins etc.). Though these substances are strongly hydrophilic, most of the water in the wall is probably held by surface tension forces in the inter-fibrillar spaces. The equivalent radius of these spaces is of the order of 10^{-7} m, but may be even smaller if the walls are cutinized. We see from Table 11.1 that pores of this size are capable of supporting water columns at least 100 m high, sufficient therefore to raise water to the tops of the tallest trees. Despite the very high tensions that are developed in the walls (of the order of 1.5×10^6 N m^{-2} or 15 bar), they will still lose water to air with a relative humidity of less than about 0.9. In other words except when the stomata are closed and/or the surrounding air is near saturation, evaporation must occur from the mesophyll cell walls.

If evaporation temporarily exceeds the supply, water is withdrawn from the walls and the menisci retreat into the inter-fibrillar spaces so setting up a considerable capillary tension. Directly, or via other cells, the cells from which evaporation occurs are in contact with the leaf xylem which is continuous with the xylem conducting systems of stem and roots and thence, via the root cortex, with water present in the surrounding soil. Provided that this pathway from the evaporating sites to the soil water is unbroken and there is sufficient water in the soil, water will be withdrawn from the latter by the tension developed in the leaves to replace that lost by evaporation and so maintain the transpiration stream.

The above explanation was proposed in principle by Dixon as long ago as 1914 in his famous cohesion theory of the ascent of sap. Cohesion refers to the ability of water columns to withstand the high tensions imposed by transpiration and opposition to this view was one of the main reasons why the theory took so long to be generally accepted. Experiments had suggested that water in glass capillaries tends to cavitate rather readily when under tension. However, it is now believed that probably because of the porous nature of the walls of the xylem elements, the water columns in plants are much more stable.

Another serious objection to the theory was based on the presumed ready disruption of the water columns by air since there is little doubt that the larger conducting elements (the vessels) frequently become filled with air; the heartwood of trees is entirely air-filled. Under the prevailing tensions therefore, it might be expected that air would spread throughout the entire conducting system and render the cohesion theory untenable. That this is not the case was shown by Scholander (1958) in his experiments on the vine. He drew attention to the occurrence at intervals along the conducting elements of cross walls perforated by fine pores. If an element is completely filled with water, these pores simply offer an increased resistance to flow. When an element is cut below a cross wall (Fig. 11.2), entering air will expand under the prevailing tension, but cannot pass beyond the cross wall, because water present in the pores is held with such high capillary forces that it cannot be forced out.

The pressure chamber (Fig. 11.3) widely used to measure tensions within plants is based on the same principle (Scholander, Hammel, Bradstreet, and Hemmington 1965). When a shoot is cut, the air that enters will spread up the stem only as far as the nearest porous barrier, whilst the sap on the other side is held in the pores with the same tension that existed in the stem

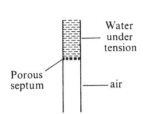

FIG. 11.2. Function of a porous membrane in preventing the spread of air in a conducting system under tension.

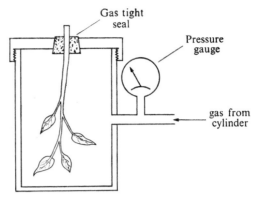

FIG. 11.3 Pressure chamber for measuring water potentials of plants; for details see text.

before it was cut. If the shoot is now placed in an airtight chamber with the cut end exposed and the pressure in the chamber is gradually raised (using a high-pressure gas supply) a stage will be reached when sap just appears at the end. The chamber pressure at which this occurs may then be assumed to correspond to the tension in the shoot before cutting. This technique has proved of immense value for the direct measurement of tensions prevailing in plants and with certain qualifications, for the evaluation of plant water potentials. Theoretical and experimental aspects of the technique are discussed in some detail by Tyree, Dainty, and Benis (1973).

Resistances to water flow

Returning to the thermodynamic concepts discussed earlier in this chapter we can regard the movement of water through the soil–plant–atmosphere continuum as occurring along a gradient of decreasing water potential (more strictly, a gradient of decreasing free energy). Typical values for an actively transpiring tree, 30 m tall with soil moisture at field capacity and the air at a relative humidity of 0·5, would be as below.

	Soil	Roots	Leaves	Air
$\psi(\text{N m}^{-2})$	-1×10^4	-2×10^5	-5×10^5	-9×10^7
Tension (bar)	0·1	2	5	900

We see that by far the biggest drop in ψ occurs between the leaf tissues and the air, corresponding to the pathway of water vapour through the stomata.

The situation is made clearer if, as originally proposed by Gradman, we adopt the analogy with Ohm's law and regard the flow of water as being 'driven' by a difference in water potential through a resistance, i.e.

$$\text{flow rate} = \frac{\Delta\psi}{R}.$$

We have already used this concept in the case of vapour flow along a concentration gradient to define a diffusion resistance (p. 115) and are extending it to gradients of water potential and liquid flow. We can now assign to difference parts of the pathway a difference in water potential ($\Delta\psi$) and a flow resistance (R). Since we are concerned mainly with the principles of this approach, we shall adopt a fairly simple model and consider only the main divisions of the water pathway; each of these can be subdivided if a more detailed analysis is required. Assuming then, a steady rate of transpiration we have:

$$\text{flow rate} = \frac{\psi_{soil}-\psi_{root}}{R_m} = \frac{\psi_{root}-\psi_{leaves}}{R_r+R_x} = \frac{\psi_{leaves}-\psi_{air}}{R_s+R_a}.$$

where the R subscripts refer to the path of liquid water through the soil to the root surface (m), across the root (r) and up the xylem to the evaporating surfaces within the leaves (x), then as water vapour through the stomata (s) and surrounding air (a).

It follows that if $\psi_{leaf}-\psi_{air}$ corresponds to the biggest drop in water potential, R_s+R_a must represent the highest resistance to flow. It is at this stage therefore that flow through the system as a whole can be most effectively controlled and this of course is achieved by regulating the stomatal opening (p. 116).

Several authors have queried the legitimacy of comparisons between resistances to vapour flow and liquid flow on the argument that the former are based on vapour concentration gradients whilst the latter are derived from hydrostatic gradients. Because of differences in the units we cannot of course equate the vapour diffusion resistances calculated earlier from concentration gradients (units s m^{-1}) with diffusion resistances calculated from water potential gradients, nor is there any obvious physical similarity between vapour flow in the leaf–air pathway and liquid flow in the xylem conducting system. Nevertheless, all flows are basically driven by differences in water potential, and it is largely a matter of practical expediency to introduce the concept of resistance as long as the theoretical limitations of this approach are appreciated.

The resistance to the flow of liquid water between leaf and root comprises a xylem resistance (R_x) and a root resistance (R_r). The resistance of the xylem conducting system has already been discussed in relation to the pressure gradients required to propel the sap through stems at certain rates (p. 34). Observations that the pressure gradient in tree stems more or less coincides with the hydrostatic gradient of 10^4 N m^{-2} per m of stem suggests that the resistance offered by the main stem at least is relatively small; this might be expected considering the vast number of conducting elements involved with flow in the stem cross-section. Roots on the other hand generally offer an appreciable resistance to water flow, mostly associated with the pathway from the epidermis to stele across the cortex, where it is under some degree of physiological control because of the Casparian strip in the endodermal cells. Thus at low temperatures or with deficient aeration, the permeability of the roots is reduced and the resistance to flow correspondingly increased. Estimates of the pressure differences required to move water through root systems at rates similar to those in transpiring plants vary considerably from about 4 to 50 bar (Kramer 1969); it is generally believed that this root resistance is responsible for the well-known lag between transpiration and absorption which frequently leads to leaf water deficits and stomatal closure during periods of rapid transpiration.

Soil moisture relations

The supply of water from a soil to a root system will be determined by the intensity of rooting and by the soil moisture content. Although soil moisture contains a certain amount of dissolved minerals, the soil water potential is dominated by its matric component. As water is removed, the larger pores are the first to empty and since their capillary attraction is relatively small, the fall in potential is initially rather small. Water remaining in the smaller pores however, is retained with much greater capillary forces and as these are progressively depleted, ψ_{soil} falls more rapidly. The precise pattern will depend on the pore space distribution; Fig. 11.4 illustrates the situation for two contrasting soil types, a heavy clay and a loam.

An important consequence of the progressive restriction of moisture to the smaller and smaller pores as the soil dries out is the rapid fall in hydraulic conductivity and hence in the resistance, R_m of the soil to water flow. This means that if flow to the roots is to be maintained at a rate sufficient to satisfy transpiration demands, the gradient in potential between the soil and root must be increased and this is only possible by a decrease in the root and therefore also in the leaf potential. As the soil dries out and its potential decreases still further, the potential of the plant as a whole must fall to a still lower level. Although to some extent, the situation may be relieved by partial stomatal closure during the day and by absorption from the soil during the night when there is no transpiration, as long as the soil continues to dry out, there will be a progressive fall in the plant water content until a stage is reached when the plant wilts, i.e. the leaf tissues lose their turgidity. At this stage, as we have seen (p. 185) the leaf water potential corresponds to the osmotic pressure of the cell sap. With loss of turgor, we would also expect the stomata to be completely closed, and though a little water may still be lost through the epidermis of the leaves (*cuticular transpiration*), there is no longer any appreciable flow of water in the system or a gradient of potential;

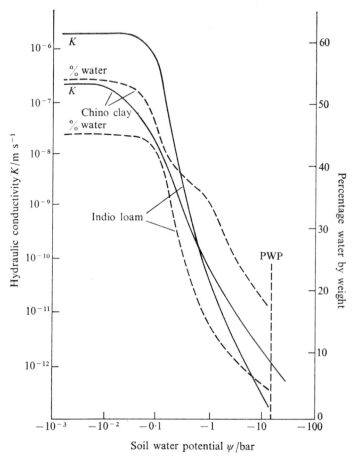

FIG. 11.4. Relationship between soil water potential (ψ) and moisture content (pecked lines) and between soil hydraulic conductivity (K) and soil water potential (continuous lines) for two contrasting soil types. PWP = permanent wilting point at $\psi \approx -15$ bars. Redrawn from the data of Gardner (1960) and Russell (1973).

under these circumstances the potentials of leaves, roots and soil would be approximately the same.

The soil moisture content (or equivalent potential) at which permanent wilting occurs is known as the *permanent wilting point* (PWP). Experiments made early in the century by Briggs and Schantz suggested that for a wide variety of species, permanent wilting occurred at more or less the same soil moisture content, later equated with a potential of about -15 bar. It was therefore concluded that permanent wilting was a function of the soil rather rather than of the crop.

The concept of the permanent wilting point as a soil constant has been criticized by several authors. It has been argued (see Slatyer 1967) that since wilting is due to loss of turgor in the leaves, when the leaf water potential corresponds to the osmotic pressure of the leaf sap, then at wilting and in the absence of a gradient, the soil water potential should be approximately the same as the leaf osmotic pressure. Since the latter ranges from about 5 to 200 bar, the soil PWP would be expected to have a similar range. Nevertheless, because for most crop plants, leaf osmotic pressures lie within a much narrower range (about 10 to 20 bar) and because the soil moisture potential changes very rapidly with even small changes in moisture content below about -0.5 bar, (Fig. 11.4), the 15 bar limit can be regarded as a soil 'constant' for most practical purposes.

Some authors have extended the concept and have defined the PWP as the lower limit for available water in the soil. Though it is generally agreed that under certain circumstances, soil water may be depleted by plants well below this level (possibly by cuticular transpiration) this extended concept is still very useful as a practical guide, if not taken too literally.

Water movement through membranes and tissues

In almost all the systems discussed so far, in order to avoid complications due to solutes, the movement of liquid water has been treated as that of pure water under a hydrostatic pressure gradient. This allowed us to derive quantitative relationships based essentially on Poiseuille's law for what may be called *mass* or *bulk flow*. The only other force that has been mentioned which also gives rise to water flow, is an osmotic gradient maintained by a difference in solute concentration, and therefore in water potential, across a semi-permeable membrane. In recent years, evidence has accumulated which suggests that water flow through such a membrane can still be regarded as mass flow. For example, it has been found that osmotic pressure gradients $(\Delta\pi)$ and hydrostatic pressure gradients (Δp) have equivalent effects on water flow through both natural and artificial membranes.

The explanation for these observations is based on the argument that if the membrane is perfectly semi-permeable, so that solute is completely excluded from the pores (assuming that pores do exist), then the pores must be entirely filled with water. Imagine such a membrane separating pure water from a salt solution as in Fig. 11.5. If only pure water occupies the pores,

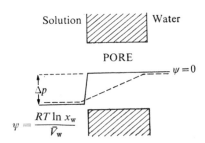

Fig. 11.5. Profile of water potential (ψ) and hydrostatic pressure (p) across a narrow pore in a membrane. After Dainty (1963).

$$\psi = \frac{RT \ln x_w}{\overline{V}_w}$$

there will be a sharp drop in water potential at the solution end of the pore. If the temperatures and external pressures on each side are the same, we have on one side, $\psi = 0$ (pure water) and on the other, $\psi = (RT/\bar{V}_w)\ln x_w$ where as in eqn (11.9), x_w is the mole fraction of water in the solution. Now we know that water flow through the membrane can be stopped and the system brought to equilibrium by applying an extra hydrostatic pressure Δp to the solution such that:

$$\Delta p + \frac{RT}{\bar{V}_w}\ln(x_w) = 0.$$

In other words, in the steady state system illustrated in Fig. 11.5 we have in effect, a hydrostatic pressure difference Δp across the membrane. Without going into further detail (see Dainty 1963) this is envisaged as giving rise to a hydrostatic pressure gradient within the pore which drives the water through by mass flow.

If we assume mass flow through pores in a membrane, its permeability should correspond to the hydraulic permeability discussed earlier (p. 29). Numerous measurements have been made based on rates of water flow into or out of various cells under known hydrostatic or osmotic pressure gradients. Published values range from about 0.2×10^{-7} cm s^{-1} for amoeba to well over 150×10^{-7} cm s^{-1} per atmosphere pressure difference for several erythrocytes (House 1974). Though the issue is still somewhat controversial, it is widely believed that the major resistance to water flow through cells is imposed by the plasmalemma (plus tonoplast in plant cells).

Water–solute interactions

Most if not all natural membranes are permeable to some extent to solutes. Since this will reduce the concentration difference across the membrane, the osmotic pressure, and therefore the flow rate, will be reduced. The ratio between the observed osmotic pressure and the theoretical value (as determined by RTc_M) gives a measure of the degree to which the solute is excluded by the membrane. This ratio defines the so-called *reflection coefficient* (σ) of the membrane and it will vary not only with the kind of membrane, but also with the nature of the solute. In the case of erythrocytes for example, σ ranges from about 0.62 for urea to 0.85 for glycerol, and for muscle tissue, from about 0.88 for urea to 1.0 (i.e. complete exclusion) for mannitol and sucrose (House 1974).

The simultaneous movement of solutes and water through a membrane may have marked mutual effects on their rates of movement. Because ions and some molecules have a number of water molecules more or less firmly attached to them (solvation shells, p. 3), solute movement is accompanied by water transport quite independently of other forces responsible for water movement in the system. In an osmotic system incorporating a membrane partially permeable to solute diffusion, the flow of water in one direction will be offset to some extent by water transported by solute particles diffusing in the opposite direction.

Electro-kinetic potentials

Another kind of interaction between water and solutes (electrolytes) which influences water movement appears in the form of a gradient in electric potential.

As the simplest example, consider a system comprising an aqueous medium in which there is a gradient of electrolyte concentration. Depending on their size and degree of solvation, different species of ion will travel at different speeds; if say, the cations diffuse faster than the anions, positive charges will tend to accumulate in the direction of flow giving rise to a gradient of electric potential, known as a *diffusion potential*. As this builds up, it will repel and slow down further movement of cations, but attract and therefore speed up anions so that, assuming that they carry equal charges, their rates of movement will soon become the same.

Now let us consider what happens when a dilute aqueous solution of an electrolyte is impelled under a small hydrostatic pressure through a narrow pore. For simplicity we shall assume an electrolyte giving rise to one species of monovalent cation and one monovalent anion. Most natural surfaces (wood, cellulose, protein, clay particles) carry a surface electrical charge; if, as is usually the case, this charge is negative, it will attract cations to form what is called an *electrical double layer* at the surface (Fig. 11.6). It has been suggested that for M-NaCl, the depth of this layer is about 100 Å, but there is no abrupt change and it is probably better to picture a more gradual falling-off in cation concentration, and an increasing anion concentration, with distance from the surface. As a result of the higher proportion of anions near the axis of the pore where the flow rate is fastest, more anions will be transported through the pore than cations. This will give rise to a difference in electric potential across the membrane, the so-called *streaming potential*; in this particular example, it will be directed (i.e. positive to negative) in the

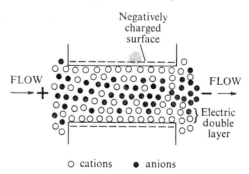

FIG. 11.6. Origin of the streaming potential across a porous membrane; $\bigcirc$ = cations, $\bullet$ = anions.

$\bigcirc$ cations　　$\bullet$ anions

same direction as the flow. Again, as this potential difference builds up it will tend to oppose further movement of anions and hence decrease the amount of water transported by the ions.

If the pores (or channels) are relatively wide, such as in the xylem or phloem conducting elements of plants, or the intercellular spaces of epithelia, the effect of the electrical double layer will be relatively small, the streaming potential likewise will be small and the effect on water flow probably negligible. On the other hand, with pores of the dimensions generally assumed to occur in membranes the effect may be appreciable. Without going into further detail, the magnitude of the streaming potential (ΔE) is linearly related to the hydrostatic pressure gradient (Δp) driving the solution through the membrane.

If, instead of applying a hydrostatic pressure to impel the solution through the membrane, a potential difference is applied across the membrane, water will flow through the membrane in the opposite direction, i.e. if positive and negative electrodes are installed as in Fig. 11.6, water will move towards the positive anode. This phenomenon, which is the converse of the streaming potential, is known as *electro-osmosis* and in this example, arises from the transport of water molecules by cations attracted towards the cathode.

Both streaming and electro-osmosis have been observed in several epithelia and single plant cells, though the situation is confused by local changes in ion concentrations near membranes in what are known as unstirred layers (see House 1974).

Irreversible thermodynamics and Onsager equations

It will be evident from the above discussion that the flow of water, especially through porous membranes, is complicated by interactions with solute flow and electrokinetic phenomena. To some extent, classical thermodynamic concepts such as were described earlier in this chapter can still be applied; however, we are no longer really concerned with equilibrium systems on which these concepts are based, but with what are known as *irreversible* or *steady state* systems. The pioneer work in this field was carried out by Onsager who devised a scheme of transport equations which have been successfully applied to biological systems. These equations are essentially phenomenological, i.e. they describe relationships at the purely macroscopic level without reference to the underlying molecular details; as long as the system does not change too rapidly, the equations may be assumed to be independent of such details.

Only the barest outline will be given here to illustrate the basic principles; details can be obtained from the now extensive literature on irreversible thermodynamics (e.g. Katchalsky and Curran 1967).

We begin by expressing the flow of water, solutes and electric current through a membrane (or tissue) in terms of their various component flows.

Assuming that the system is isothermal we can write:

$$\text{water flow} = \textit{hydraulic flow} + \text{solute linked flow}$$
$$+ \text{electro-osmotic flow}$$
$$\text{solute flow} = \text{ultrafiltration} + \textit{diffusion} + \text{electrophoresis}$$
$$\text{electric current flow} = \text{streaming current} + \text{diffusion current}$$
$$+ \textit{electric potential current.}$$

Each of these components refers to a specific driving force; those in italics refer to so-called *conjugate* forces which are directly concerned with the flow of that quantity, whilst the others refer to forces linked or coupled to the flow of the other quantities. An important point to note is that each of these coupled forces is the converse of another. We have already seen that electro-osmotic flow is the converse of streaming potential flow; similarly, solute linked flow which refers to water transported by solutes diffusing through the membrane is the converse of ultrafiltration which is the transport of solutes by water flowing through the membrane, whilst electrophoresis, the transport of charged particles in an electric field is the converse of the diffusion current resulting from the transport of electric charges by ions.

The next stage is to express each of the component flows in terms of the product of the force responsible for that flow and a corresponding coefficient. Using the notation conventional to Onsager equations in which flow rates (fluxes) are denoted by J and coefficients by L (both identified by appropriate subscripts) we have:

for water; $\quad J_v(\text{m}^3 \text{ m}^{-2} \text{ s}^{-1}) = L_p \, \Delta p + L_{pD} \, \Delta\pi + L_{pE} \, \Delta E$

for solutes; $\quad J_D(\text{mol m}^{-2} \text{ s}^{-1}) = L_{Dp} \, \Delta p + L_D \, \Delta\pi + L_{DE} \, \Delta E$

for current; $\quad J_E(\text{A}) = L_{Ep} \, \Delta p + L_{ED} \, \Delta\pi + L_E \, \Delta E.$

We see that Δp, the pressure gradient across the membrane which is the conjugate force for water flow, is also the coupled force for solute flow (coefficient L_{Dp}) and for current flow (coefficient L_{Ep}). Similarly, the osmotic gradient $\Delta\pi$ which is the conjugate force for solute flow (being determined by the concentration difference across the membrane) is the coupled force for osmotic water flow (L_{pD}) and current flow (L_{ED}). The electric potential difference across the membrane ΔE, representing the conjugate force for current flow, is the coupled force for electro-osmotic and electrophoretic flows. A significant feature of these equations is that related coupling coefficients, L_{pD} and L_{Dp}; L_{pE} and L_{Ep}; L_{DE} and L_{ED} are the same; for a complete solution therefore, only 6 coefficients are required instead of 9.

Before the above equations can be applied however, a certain amount of manipulation is necessary to convert the component forces and coefficients into measurable quantities. No attempt will be made to describe these manipulations, but one working example of a water flux equation is given below to illustrate the kind of measurements that need to be made:

$$J_v = L_p(\Delta p - \Delta \pi) + L_p(1 - \sigma)\,\Delta \pi_s - L_p(P_E I)/K.$$

It will be noted that the component flows have been so arranged that the coupling coefficients are all expressed in terms of the same hydraulic permeability L_p.

The hydraulic permeability is determined from the volume flux of water (J_v) under a given pressure gradient ($\Delta p - \Delta \pi$) when $\Delta \pi_s$ and I are held at zero:

$$L_p = \frac{J_v}{\Delta p - \Delta \pi}.$$

This is the same hydraulic permeability described in Chapter 3. $\Delta \pi_s$ refers to the theoretical osmotic pressure difference as given by $RT\,\Delta c_s$ where Δc_s is the difference in *permeable* solute concentration (in moles) across the membrane.

$\Delta \pi$ refers to the osmotic pressure due to Δc_s plus any extra due to impermeable solutes.

The reflection coefficient σ is determined from the solute flux associated with a given volume flow driven say, by a known hydrostatic gradient, when both $\Delta \pi_s$ and I are zero. If $\bar{c}_s$ is the mean solute concentration on both sides of the membrane,

$$\bar{c}_s(1 - \sigma) = \frac{J_s}{J_v}.$$

The electro-osmotic pressure P_E is obtained from the ratio of the pressure driving force $\Delta p - \Delta \pi$ to the potential difference ΔE developed across the membrane when J_v and $\Delta \pi_s$ are zero. If the electric current I is recorded when an electrical potential E is applied across the membrane (both $\Delta p - \Delta \pi$ and $\Delta \pi_s$ being zero) the *specific conductance K* is obtained from

$$K = \frac{I}{E}.$$

Spanner (1970) has suggested that electro-osmosis is responsible for phloem transport across the sieve plates, but generally speaking, transport by electro-osmosis is much less efficient than that by pressure flow. Tyree (1971) has calculated that the electrical gradients required to produce the same water fluxes as a pressure gradient of 10^4 N m^{-2} m^{-1} in petiole and stem xylem would be $7 \cdot 8 \times 10^3$ and 13×10^3 V m^{-1} respectively, thus considerably higher than the potentials normally found in living trees (of the order of 0·2 to 0·4 V m^{-1}). In the phloem, a potential gradient of $1 \cdot 8 \times 10^3$ V m^{-1} would be needed to produce the same volume flux as a pressure gradient of 10^5 N m^{-2} m^{-1}, whereas the potentials normally observed are of the order of 0·05 to 0·10 V m^{-1}. On the other hand, over short distances such as within meristematic tissues, in which potentials of 50 mV per cm have been recorded, electrical gradients may be of much greater importance.

The dissipation function and energy relations

The Onsager equations can be used to derive a *dissipation function* (a measure of the dissipation of free energy, and therefore, with certain restrictions, of the power consumption) when water is driven through a porous system. This follows from the fact that provided that the system is not too far from equilibrium, the product of a flux and a force gives a direct measure of the work done. Tyree (1971) discusses some applications of the Onsager equations to the energy relations of water transport in trees; he concludes that by far the greatest dissipation of energy occurs in the vapour phase, first in the evaporation of water and then in the diffusion of water vapour from the internal leaf tissues into the surrounding air. By contrast, the energy consumed in the cell walls of the xylem and that required to lift water to a height of 10 m were relatively small. These conclusions are in accord with those discussed earlier using a different approach.

Water diffusion in cells

Underlying the above account of membrane transport is the assumption that water and solutes share a common pathway through 'pores' in the membrane. Though several features of membrane behaviour can apparently be explained in this way, the assumption has several weaknesses. For example, the widely-used concept of a hydraulic permeability assumes that Poiseuille's law applies, yet measurements on a variety of natural membranes, using methods based on this assumption, indicate equivalent pore radii mostly of the order of 5 to 10 Å. Few if any attempts appear to have been made to reconcile these findings with the presumed blocking of the pores by ordered water (p. 27).

Even if pores of a kind do exist in biological membranes, it is most unlikely that water and solutes move exclusively in them; some diffusion must occur through the membrane and certainly in the cytoplasm. Using labelled water, several measurements have been made of water diffusion into and out of cells and allowing for considerable experimental uncertainties, published values for diffusional permeabilities cover a wide range, from 0.2×10^{-4} (amoeba) to possibly over 20×10^{-4} cm s^{-1} for certain erythrocytes.

A comparison of the diffusion permeability of a cell (P_D) with its hydraulic permeability (L_p) in terms of the ratio ($L_p RT / \bar{V} P_D$) indicates a wide range of values, but mostly lying between 1 and 5. Since the same ratio for non-porous artificial membranes is between 1 and 2, it is possible that many biological membranes are indeed, non-porous.

The problems of membrane structure and transport are far too complex to permit any further meaningful discussion here. There are several excellent texts and reviews on the subject, including House (1974), from which most of the above mentioned data have been taken.

Appendices

Appendix 1: Laminar flow through circular tubes (Ch. 2)

THE following assumes that the tube is of uniform radius, that its walls are rigid, and that a sufficient length has been traversed so that the boundary layer is fully developed.

Consider a cylindrical element of fluid, symmetrical about the tube axis, of length l and radius y, moving at uniform speed u (Fig. A.1). Because of

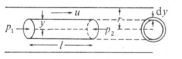

FIG. A.1

viscous forces acting along the surface of this element and opposing its motion, a certain driving force is necessary to keep it moving and as a result of the energy consumed, there will be a regular fall in pressure along the tube.

If the pressures at the two ends of the element are p_1 and p_2, they will exert forces $p_1\pi y^2$ and $p_2\pi y^2$; since they act in opposite directions (fluid pressure on a surface always acts normal to the surface), the net driving force on the element will be $\pi y^2(p_1-p_2)$.

The shear stress τ opposing the movement of the element acts parallel to its outer surface, of area $2\pi yl$. Hence the total drag on the element will be $-2\pi yl\tau$ (negative because it acts in the direction opposite to that of flow).

Since the element moves at constant speed, i.e. there is no acceleration, then according to Newton's second law the forces acting on it must balance:

$$\pi y^2(p_1-p_2) = -2\pi yl\tau, \qquad (A.1)$$

whence

$$\tau = -\frac{(p_1-p_2)y}{2l} = -\frac{y\,\Delta p}{2l}, \quad \text{as text eqn (2.1}a).$$

where $\Delta p = p_1-p_2$.

Eqn (A.1) is of general application to flow in tubes irrespective of the nature of flow.

With laminar flow, text eqn (1.1) applies, i.e. $\tau = \eta(du/dy)$. Substituting for τ in eqn (A.1) gives

$$\frac{du}{dy} = -\frac{y\,\Delta p}{2l\eta}. \qquad (A.2)$$

Integrating eqn (A.2) gives

$$u = -\frac{y^2\,\Delta p}{4\eta l} + \text{constant}. \qquad (A.3)$$

Since fluid velocity at the wall is zero (no slip), $u = 0$ when $y = r$, the radius of the tube. Substituting this in eqn (A.3) gives

$$u = \frac{(r^2 - y^2)\,\Delta p}{4\eta l}.$$ (A.4)

Eqn (A.4) represents the laminar velocity profile across the tube and corresponds to a parabola.

Now consider an elementary annulus around the cylindrical element of thickness dy (Fig. A.1). The cross-sectional area of this annulus is $2\pi y\,dy$, so that with a velocity u, the volume of fluid V flowing through it in unit time will be $2\pi u y\,dy$. Hence, q_v the total volume flow rate across the whole of the tube can be obtained by integrating V from $y = 0$ to $y = r$:

$$q_v = \int_0^r 2\pi u y\,dy.$$ (A.5)

Substituting for u from eqn (A.4) gives:

$$q_v = \int_0^r \frac{2\pi(r^2 - y^2)\,\Delta p}{4\eta l}\, y\,dy$$

whence,

$$q_v = \frac{\pi r^4 \Delta p}{8\eta l},$$

as text eqn (2.2).

The maximum velocity u_{max} occurs at the axis of the tube where $y = 0$. Hence, from eqn (A.4),

$$u_{max} = \frac{r^2 \Delta p}{4\eta l},$$

as text eqn (2.3).

Appendix 2: Turbulent flow in tubes (Ch. 2)

Assume that the velocity profile is represented by text eqn (2.12):

$$u = 8 \cdot 77 u_* \left(\frac{z u_*}{\nu}\right)^{\frac{1}{7}},$$ (A.6)

where $z =$ distance from wall. At the axis ($z = r$), $u = u_{max}$, therefore,

$$u_{max} = 8 \cdot 77 u_* \left(\frac{r u_*}{\nu}\right)^{\frac{1}{7}}.$$ (A.7)

Dividing eqn (A.6) by (A.7) gives:

$$u = u_{max}\left(\frac{z}{r}\right)^{\frac{1}{7}}.$$ (A.8)

Consider an annulus of radius $y(= r - z)$ and thickness dz symmetrical about the axis (as Fig. A.1, but with dz instead of dy). The area of this annulus $= 2\pi y\,dz = 2\pi(r - z)\,dz$ and if the velocity of flow $= u$, the volume rate of flow through it $= 2\pi u(r-z)\,dz$. The volume rate of flow q_v through the whole tube will then be

$$q_v = \int_0^r 2\pi(r-z)u\,dz = \pi r^2 u_m, \tag{A.9}$$

where u_m = average velocity in tube.

Substituting for u from eqn (A.8) gives

$$\int_0^r 2\pi(r-z)u_{max}\left(\frac{z}{r}\right)^{\frac{1}{7}} dz = \pi r^2 u_m,$$

whence

$$u_m = \frac{2u_{max}}{r^{\frac{15}{7}}}\int_0^r (r-z)(z)^{\frac{1}{7}}\,dz \tag{A.10}$$

Integrating eqn (A.10) gives

$$u_m = 0.817u_{max}. \tag{A.11}$$

Substituting $u_* = \sqrt{(\tau_0/\rho)}$ into eqn (A.7) gives

$$u_{max} = 8.77\,\frac{\tau_0^{\frac{4}{7}}r^{\frac{1}{7}}}{\rho^{\frac{4}{7}}\nu^{\frac{1}{7}}} = \frac{u_m}{0.817},$$

whence

$$\tau_0 = \frac{u_m^{\frac{7}{4}}\rho\nu^{\frac{1}{4}}}{(7.15)^{\frac{7}{4}}r^{\frac{1}{4}}}.$$

Putting $d = 2r$ and $(Re) = u_m d/\nu$ gives

$$\tau_0 = 0.076\,\frac{\rho u_m^2}{2}\,(Re)^{-\frac{1}{4}}.$$

With a minor adjustment to fit the experimental data better, we obtain

$$\tau_0 = 0.0791\,\frac{\rho u_m^2}{2}\,(Re)^{-\frac{1}{4}},$$

as in text eqn (2.13). Putting $\Delta p = 2l\tau_0/r$ (from text eqn (2.1b)),

$$(Re) = \frac{2u_m r}{\nu} \quad\text{and}\quad u_m = \frac{q_v}{\pi r^2}$$

gives

$$\Delta p \propto \frac{l\rho^{\frac{3}{4}}\eta^{\frac{1}{4}}q_v^{\frac{7}{4}}}{r^{\frac{19}{4}}},$$

i.e. with turbulent flow, Δp is proportional to $q_v^{\frac{7}{4}}$.

Appendix 3: Depth of the laminar sub-layer (Ch. 2)

Assuming $zu_*/\nu = \delta_s u_*/\nu = 10$, then $\delta_s = 10\nu/u_*$. From text eqn (2.13)

$$\frac{\tau_0}{\rho} = 0.0791\,\frac{u_m^2}{2}\,(Re)^{-\frac{1}{4}}$$

whence

$$u_* = \sqrt{\left(\frac{\tau_0}{\rho}\right)} \approx 0\cdot 2u_{\mathrm{m}}(Re)^{-\frac{1}{8}}$$

and

$$\delta_{\mathrm{s}} \approx \frac{10v(Re)^{\frac{1}{8}}}{0\cdot 2u_{\mathrm{m}}} \approx \frac{50vu_{\mathrm{m}}^{\frac{1}{8}}d^{\frac{1}{8}}}{u_{\mathrm{m}}v^{\frac{1}{8}}} \approx \frac{50dv^{\frac{7}{8}}}{u_{\mathrm{m}}^{\frac{7}{8}}d^{\frac{7}{8}}} \approx \frac{50d}{(Re)^{\frac{7}{8}}}.$$

Appendix 4: Compressible gas flow (Ch. 3)

Darcy's law for incompressible fluids is expressed in text eqn (3.4) as

$$\phi_{\mathrm{v}} = \frac{k\,\Delta p}{\eta l}. \tag{A.12}$$

For compressible gases this is written in the derivative form

$$\phi_{\mathrm{v}} = -\frac{k}{\eta}\cdot\frac{\mathrm{d}p}{\mathrm{d}l} \tag{A.13}$$

(negative because p decreases with increasing l). According to the universal gas law, $pV = NRT$, hence the product of the volume flux ϕ_{v} and the pressure p at any point in the medium must remain constant if the temperature T is constant. Multiplying both sides of eqn (A.13) by p gives

$$p\phi_{\mathrm{v}} = -\frac{kp}{\eta}\cdot\frac{\mathrm{d}p}{\mathrm{d}l} \quad \text{or} \quad (p\phi_{\mathrm{v}})\,\mathrm{d}l = -\frac{k}{\eta}\,p.\mathrm{d}p. \tag{A.14}$$

If along a length of medium l, the pressure drops from p_1 to p_2, we can integrate eqn (A.14) for l from 0 to l, and for p, from p_1 to p_2:

$$\int_0^l (p\phi_{\mathrm{v}})\,\mathrm{d}l = -\int_{p_1}^{p_2} \frac{k}{\eta}\,p.\mathrm{d}p. \tag{A.15}$$

Assuming that k and η remain constant, eqn (A.15) gives

$$(p\phi_{\mathrm{v}})l = \frac{k}{2\eta}(p_1^2 - p_2^2) \quad \text{or} \quad \phi_{\mathrm{v}} = \frac{k}{\eta}\cdot\frac{(p_1+p_2)(p_1-p_2)}{2pl}. \tag{A.16}$$

Now $(p_1+p_2)/2$ corresponds to the average pressure $\bar{p}$ over the length l of medium and p_1-p_2 to the pressure drop Δp across the medium. Hence we can rewrite eqn (A.16) in the form

$$\phi_{\mathrm{v}} = \frac{k}{\eta}\cdot\frac{\bar{p}}{p}\cdot\frac{\Delta p}{l},$$

as in text eqn (3.6).

Appendix 5: Derivation of the Kozeny–Carman equation (Ch. 3)

A general equation for laminar flow in a non-circular tube of length l is

$$\frac{q_{\mathrm{v}}}{A'} = \frac{m^2\,\Delta p}{k_0\eta l}, \tag{A.17}$$

where A' is the cross-sectional area of the tube, m the hydraulic radius is cross-sectional area divided by tube perimeter, and k_0 a factor depending on the shape of the tube. It follows that in a medium with uniform pores, m is the pore volume divided by the total internal surface area of pores and A' is the total pore area in the cross-section.

If the pores occupy a fraction ϵ of the total volume and they are uniformly distributed, then the total cross-sectional area of the medium $(A) = A'/\epsilon$; also, if s is total internal surface area divided by total volume of medium $m = \epsilon/s$. Substituting these in eqn (A.17) gives

$$\frac{q_v}{\epsilon A} = \frac{\epsilon^2 \Delta p}{s^2 k_0 \eta l}, \qquad \text{whence} \qquad \frac{q_v}{A} = \frac{\epsilon^3 \Delta p}{s^2 k_0 \eta l}. \tag{A.18}$$

In a more realistic porous medium the pores would not be expected to run parallel to the flowpath, but to take a more tortuous, longer pathway through the medium. To allow for this, l is multiplied by a tortuosity factor T. If we write $c = k_0 T$, then

$$\frac{q_v}{A} = \frac{\epsilon^3 \Delta p}{s^2 c \eta l},$$

as in text eqn (3.10).

For the parallel-tube model (Fig. 3.2), with N pores in the cross-section, $\epsilon = N\pi r^2/A$; $k_0 = 2$; $T = 1$ (whence $c = 2$); and $s = 2N\pi r/A$. Substituting these values in text eqn (3.10) gives

$$q_v = \frac{N\pi r^4 \Delta p}{8\eta l},$$

which corresponds to Poiseuille's formula for N circular tubes in parallel (text eqn 3.7).

Appendix 6(a): Laminar flow over a flat plate (Ch. 4)

In text Fig. 4.1, CC' depicts the side of a plane section of the laminar boundary layer perpendicular to the plate at a distance l from 0, the leading edge. If the fluid velocity at height z in this plane is u and its density is ρ, the mass of fluid passing through an element of height dz in unit time $= \rho u \, dz$ (for a plate of unit width) and its momentum parallel to the surface will be $\rho u^2 \, dz$.

Outside the boundary layer, the velocity of this same mass of fluid was U and its momentum $\rho u U \, dz$. Hence the loss of momentum in the boundary layer $= \rho u U \, dz - \rho u \, dz = \rho u (U-u) \, dz$ and the total loss of momentum over the depth δ_L of boundary layer at C is $\int_0^{\delta_L} \rho u (U-u) \, dz$.

This loss of momentum must equal the drag force F_{DL} exerted by the plate on the fluid over the length l; therefore

$$F_{DL} = \int_0^{\delta_L} \rho u (U-u) \, dz. \tag{A.19}$$

The simplest solution to eqn (A.19) is obtained by assuming that within the boundary layer, the shear stress τ decreases linearly with distance from the surface, from τ_0 at $z = 0$ to zero at $z = \delta_L$; i.e.

$$\tau = \tau_0\left(1 - \frac{z}{\delta_L}\right). \tag{A.20}$$

From text eqn (1.1) for laminar flow, $\tau = \eta(du/dz)$. Therefore, eqn (A.20) can be written

$$\frac{du}{dz} = \frac{\tau_0}{\eta}\left(1 - \frac{z}{\delta_L}\right). \tag{A.21}$$

Integrating eqn (A.21) from $z = 0$ to $z = \delta_L$ gives

$$u = \frac{\tau_0}{\eta}\left(z - \frac{z^2}{2\delta_L}\right) \tag{A.22}$$

(since $u = 0$ at $z = 0$, the integration constant is zero).

By definition, $u = U$ when $z = \delta_L$. Substituting in eqn (A.22) and re-arranging gives

$$\tau_0 = \frac{2\eta U}{\delta_L}. \tag{A.23}$$

Substituting for τ_0 from eqn (A.23) into eqn (A.22) gives

$$u = \frac{2U}{\delta_L}\left(z - \frac{z^2}{2\delta_L}\right). \tag{A.24}$$

Hence the velocity profile is parabolic in form. To solve eqn (A.19) substitute for u from eqn (A.24). On integrating and simplifying this gives

$$F_{DL} = \frac{2}{15}\rho U^2 \delta_L. \tag{A.25}$$

As defined in eqn (A.19), F_{DL} corresponds to the skin friction τ_0 integrated over the length l of plate of unit width, i.e.

$$F_{DL} = \int_0^l \tau_0 \, dl. \tag{A.26}$$

Substituting for τ_0 from eqn (A.23) gives

$$F_{DL} = \int_0^l \frac{2\eta U}{\delta_L} \, dl \tag{A.27}$$

(n.b. δ_L is a function of l).

From eqns (A.27) and (A.25),

$$\int_0^l \frac{2\eta U}{\delta_L} \cdot dl = \frac{2}{15} \rho U^2 \delta_L \qquad (A.28)$$

giving:

$$\frac{\eta}{\delta_L} = \frac{\rho U}{15} \frac{d(\delta_L)}{dl} \qquad (A.28)$$

whence, by separation of the variables and integrating (with $\delta_L = 0$ when $l = 0$), we obtain:

$$\delta_L^2 = \frac{30\eta l}{\rho U} \quad \text{and} \quad \delta_L = 5 \cdot 48 \sqrt{\left(\frac{\eta l}{\rho U}\right)} = 5 \cdot 48 l (Re_l)^{-\frac{1}{2}}, \qquad (A.29)$$

as text eqn (4.1) with a minor adjustment to the factor. From eqns (A.23) and (A.29),

$$\tau_0 = \frac{0 \cdot 365 \eta U (Re_l)^{\frac{1}{2}}}{l} = \frac{0 \cdot 365 \rho U^2 \eta (Re_l)^{\frac{1}{2}}}{U l \rho} = 0 \cdot 365 \rho U^2 (Re_l)^{-\frac{1}{2}}.$$

A more exact solution gives

$$\tau_0 = 0 \cdot 332 \rho U^2 (Re_l)^{-\frac{1}{2}} \text{ as text eqn (4.2)}$$

Substituting for τ_0 in eqn (A.26) gives

$$F_{DL} = 0 \cdot 332 \rho U^2 \int_0^l (Re_l)^{-\frac{1}{2}} dl = 0 \cdot 664 \rho U^2 l (Re_l)^{-\frac{1}{2}}$$

as text eqn (4.3)

Appendix 6(b): Turbulent flow over a flat plate (Ch. 4)

As for laminar flow (eqn (A.19)), the total drag F_{DT} of the area of plate occupied by the turbulent boundary layer (depth δ_T) is equated to the loss of momentum of fluid in the boundary layer:

$$F_{DT} = \int_{z=0}^{z=\delta_T} \rho u (U-u) \, dz. \qquad (A.30)$$

For turbulent flow in pipes, the velocity profile well within the boundary layer can be represented by eqn (A.8):

$$\frac{u}{u_{max}} = \left(\frac{z}{r}\right)^{\frac{1}{7}}. \qquad (A.31)$$

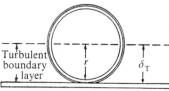

FIG. A.2

Turbulent boundary layer

We can apply this to turbulent flow over a plane surface if we assume that the boundary layer in a pipe corresponds to that over a plate which has been wrapped around an axis distant δ_T from the plate (Fig. A.2). It follows that δ_T corresponds to the tube radius r and the axial velocity u_{max} to the free stream velocity U, so that for our plate we have

$$\frac{u}{U} = \left(\frac{z}{\delta_T}\right)^{\frac{1}{7}}, \tag{A.32}$$

where $u = $ velocity at height z above the surface.

Substituting for u from eqn (A.32) into eqn (A.30) and integrating gives

$$F_{DT} = \frac{7}{72}\rho U^2 \delta_T. \tag{A.33}$$

Again as for laminar flow, F_{DT} is the skin friction τ_0 integrated over the area of plate of length l_T and unit width occupied by the turbulent boundary layer,

$$F_{DT} = \int_0^{l_T} \tau_0 \, dl = \frac{7}{72}\rho U^2 \delta_T. \tag{A.34}$$

In contrast to laminar flow, no simple relationship can be assumed between τ_0 and δ_T. Instead we can use the empirical expression for turbulent drag in pipes given by text eqn (2.13):

$$\tau_0 = 0.0791\rho\frac{u_m^2}{2}\left(\frac{\nu}{u_m d}\right)^{\frac{1}{4}}. \tag{A.35}$$

By analogy with the unrolled pipe, $d = 2\delta_T$, and since $u_m \approx 0.8u_{max}$ (Appendix 2), $u_m \approx 0.8U$, the free stream velocity. Substituting for d and u_m in eqn (A.35) gives

$$\tau_0 = 0.023\rho U^2\left(\frac{\nu}{U\delta_T}\right)^{\frac{1}{4}}, \tag{A.36}$$

as text eqn (4.5)
and substituting for τ_0 from eqn (A.36) into eqn (A.34) gives

$$F_{DT} = \int_0^{l_T} 0.023\rho U^2\left(\frac{\nu}{U\,\delta_T}\right)^{\frac{1}{4}} dl = \frac{7}{72}\rho U^2 \delta_T,$$

whence

$$\delta_T = 0.376 l_T (Re_{l_T})^{-\frac{1}{5}}, \tag{A.37}$$

as text eqn (4.4). Finally, substituting for δ_T from eqn (A.37) into eqn (A.34) gives

$$F_{DT} = 0.0366\rho U^2 l_T (Re_{l_T})^{-\frac{1}{5}},$$

as text eqn (4.7).

Appendix 7: Stokes' law (Ch. 5)

When a spherical particle falls through a fluid it will eventually attain a steady velocity U (the terminal velocity) when the gravitational force is balanced by forces opposing its motion, i.e. when the weight of the particle is balanced by the upthrust plus drag force.

The weight of the particle equals volume $(\pi d^3/6) \times$ density $(\rho_s) \times$ acceleration due to gravity (g), where d is the particle diameter.

The upthrust equals the weight of fluid displaced, which is equal to the volume of particle $(\pi d^3/6) \times$ density of fluid $(\rho_t) \times g$.

At Reynolds numbers below about 1, the drag force F_D is given by

$$F_D = 3\pi d\eta U \text{ (Stokes' law),}$$

where η is the dynamic viscosity of fluid.

At equilibrium,

$$\frac{\pi d^3}{6} \rho_s g = \frac{\pi d^3}{6} \rho_t g + 3\pi d\eta U,$$

whence

$$U = \frac{d^2 g(\rho_s - \rho_t)}{18\eta} \quad \text{and} \quad \eta = \frac{d^2 g(\rho_s - \rho_t)}{18U}.$$

If ρ_s, ρ_t, and η are known, measurement of the terminal velocity U (rate of settling) gives a measure of the particle size in terms of the equivalent diameter of a sphere. The same applies to liquid droplets falling in a gas.

Appendix 8: Efficiency of rowing action (Ch. 6)

Fig. A.3 illustrates the action of a pair of 'oars' propelling a body with a steady speed U. In the power stroke (a), if the speed of the oars relative to the body is u, their speed relative to the water will be $u-U$, and if the drag force on each of N oars is f, the total power expended will be $Nf(u-U)$.

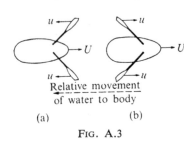

Relative movement
of water to body

(a)   (b)

Fig. A.3

On the return stroke (b) with the oars moving at the same speed relative to the body, their speed relative to the water will be $u+U$, and if the drag on each oar is f', the power expended for N oars will be $Nf'(u+U)$.

If the drag on the body itself is F, then the power expended, i.e. the mechanical propulsive power, will be FU. The total power expended during one complete cycle will be

$$Nf(u-U) + Nf'(u+U) + FU. \tag{A.38}$$

For each rowing cycle, the net propelling force exerted by each oar is $f-f'$ and for N oars will be $N(f-f')$. If the body moves at uniform speed, the net propelling force just balances the drag on the body so that

$$F = N(f-f'). \tag{A.39}$$

Substituting for F in eqn (A.38) gives the total power as

$$Nf(u-U)+Nf'(u+U)+NU(f-f') = Nu(f+f'). \qquad (A.40)$$

The mechanical propulsive power (as above) $= FU$. Substituting for F from eqn (A.39) gives the mechanical propulsive power as

$$NU(f-f').$$

Therefore the efficiency of propulsion ($=$ mechanical power/total power)

$$= \frac{NU(f-f')}{Nu(f+f')} = \frac{U(f-f')}{u(f+f')}. \qquad (A.41)$$

It follows that the efficiency will be increased by reducing u, the speed of the oars relative to U, the speed of the body, and by decreasing f', the drag on the return stroke (such as by feathering). If f is large compared with f', e.g. by using large blades for the power stroke, f' can be ignored so that the efficiency will then be determined only by U/u.

At the very low (Re) at which protozoan cilia operate, the drag forces f and f' are entirely viscous in origin and will then be proportional to the stroke velocity (see p. 53).

Appendix 9: Distance travelled by gliding fish (Ch. 6)

According to Newton's second law of motion, the drag force F on the fish at any point during the glide will be equal to the rate of loss of momentum $-m(dU/dt)$, where m is the mass of the fish and dU/dt the rate at which the velocity changes at that point. Assuming that F is proportional to the square of the velocity, i.e. $F = kU^2$, then

$$kU^2 = -m \frac{dU}{dt} \qquad \text{or} \qquad \frac{1}{U^2} dU = -\frac{k}{m} dt. \qquad (A.42)$$

Integrating eqn (A.42) gives

$$\frac{1}{U} = \frac{kt}{m} + \text{constant}. \qquad (A.43)$$

If $U = U_0$ when $t = 0$, i.e. U_0 is the velocity of the fish at the beginning of the glide, then the constant $= 1/U_0$, and eqn (A.43) can be written

$$\frac{1}{U} - \frac{1}{U_0} = \frac{kt}{m}. \qquad (A.44)$$

If the fish travels a distance ds in time dt when its velocity $= U$, then

$$ds = U\,dt.$$

Substituting for dt in eqn (A.42) gives

$$kU^2 = -mU \frac{dU}{ds} \qquad \text{or} \qquad \frac{1}{U} dU = -\frac{k}{m} ds. \qquad (A.45)$$

Integrating then gives

$$\ln U = -\frac{ks}{m} + \text{constant.} \tag{A.46}$$

If $U = U_0$ when $s = 0$, the constant $= \ln U_0$ and eqn (A.46) becomes

$$\ln\left(\frac{U}{U_0}\right) = -\frac{ks}{m} \quad \text{or} \quad \ln\left(\frac{U_0}{U}\right) = \frac{ks}{m}. \tag{A.47}$$

Substituting for $1/U$ from eqn (A.44) into eqn (A.47) gives

$$\ln\left\{U_0\left(\frac{kt}{m} + \frac{1}{U_0}\right)\right\} = \frac{ks}{m},$$

whence

$$s = \frac{m}{k}\ln\left(\frac{U_0 kt}{m} + 1\right),$$

as in text eqn (6.2).

Appendix 10: Speed for minimum drag (Ch. 6)

From text eqn (6.6), induced drag is given by

$$\frac{1}{\pi} \cdot \frac{1}{\frac{1}{2}\rho U^2}\left(\frac{\text{lift}}{b}\right)^2, \tag{A.48}$$

where b is span. From text eqn (6.5), lift force,

$$F_L = C_L \cdot \tfrac{1}{2}\rho U^2 A, \tag{A.49}$$

therefore, induced drag is given by

$$\frac{1}{\pi} \cdot \frac{1}{\frac{1}{2}\rho U^2} \cdot \frac{C_L^2(\frac{1}{2}\rho U^2 A)^2}{b^2} = \frac{1}{\pi}\frac{\frac{1}{2}\rho U^2 A^2 C_L^2}{b^2}.$$

Putting $k = A/\pi b^2$ gives induced drag as

$$kC_L^2 \tfrac{1}{2}\rho U^2 A. \tag{A.50}$$

By definition, the coefficient for *body drag+profile drag* at zero lift $= C_{D0}$. Therefore,

$$\text{body drag+profile drag} = C_{D0}\tfrac{1}{2}\rho U^2 A, \tag{A.51}$$

and the *total drag* $F_D = $ *body drag+profile drag+induced drag*, becomes

$$F_D = C_{D0}\tfrac{1}{2}\rho U^2 A + kC_L^2\tfrac{1}{2}\rho U^2 A. \tag{A.52}$$

F_D can also be expressed in terms of a total drag coefficient C_D, whence $F_D = C_D\tfrac{1}{2}\rho U^2 A$ and eqn (A.52) becomes

$$C_D = C_{D0} + kC_L^2. \tag{A.53}$$

In steady level flight, the lift force F_L must balance the body weight W, i.e. $F_L = W$. Expressing F_D as $F_L(F_D/F_L)$, we have

$$F_D = W\frac{F_D}{F_L} = W\frac{C_D}{C_L}, \tag{A.54}$$

Hence for minimum drag, C_D/C_L must be a minimum. Substituting for C_D from eqn (A.53), this means that $(C_{Do}+kC_L^2)/C_L$ must be a minimum. This minimum can be obtained by differentiating the above with respect to C_L and putting the differential $= 0$. This gives

$$(C_{Do}+kC_L^2)-C_L(2kC_L) = 0, \tag{A.55}$$

whence $C_{Do} = kC_L^2$. Since $C_{Do} = $ *body drag+profile* drag/$\frac{1}{2}\rho U^2 A$, and from eqn (A.50) $kC_L^2 = $ *induced* drag/$\frac{1}{2}\rho U^2 A$, then the *minimum* drag occurs when the *body drag+profile* drag $= $ *induced* drag.

Putting C_{Lmd} as the lift coefficient at minimum drag, we have from eqn (A.55),

$$C_{Do} = k(C_{Lmd})^2 \quad \text{or} \quad C_{Lmd} = \sqrt{\left(\frac{C_{Do}}{k}\right)} \tag{A.56}$$

and since the lift force $= $ body weight $W = C_{Lmd}\frac{1}{2}\rho(U_{md})^2 A$, where U_{md} is the speed for minimum drag, then

$$U_{md} = \sqrt{\left(\frac{W}{\frac{1}{2}\rho A C_{Lmd}}\right)} = \left(\frac{W}{\frac{1}{2}\rho A}\right)^{\frac{1}{2}}\left(\frac{k}{C_{Do}}\right)^{\frac{1}{4}},$$

as in text eqn (6.7).

Appendix 11: Minimum power relations (Ch. 6)

If F_D is the total drag force and U the speed, the power output

$$P = F_D U. \tag{A.57}$$

As in eqn (A.52)

$$F_D = C_{Do}\frac{1}{2}\rho U^2 A+kC_L^2\frac{1}{2}\rho U^2 A, \tag{A.58}$$

and since the lift force $= $ body weight $W = C_L\frac{1}{2}\rho U^2 A$, then $C_L = W/\frac{1}{2}\rho U^2 A$. Substituting for C_L in eqn (A.58) gives

$$F_D = C_{Do}\frac{1}{2}\rho U^2 A+kW^2/\frac{1}{2}\rho U^2 A,$$

therefore, power is given by

$$P = C_{Do}\frac{1}{2}\rho U^3 A+kW^2/\frac{1}{2}\rho UA = xU^3+y/U, \tag{A.59}$$

where $x = C_{Do}\frac{1}{2}\rho A$ and $y = kW^2/\frac{1}{2}\rho A$. These quantities are plotted in text Fig. 6.15. As in eqn (A.54) we can write $F_D = W(C_D/C_L)$, so

$$P = WU\frac{C_D}{C_L}. \tag{A.60}$$

Since the lift force $F_{\text{L}} = C_{\text{L}}\frac{1}{2}\rho U^2 A = $ body weight W, we can also write

$$W = C_{\text{L}}\tfrac{1}{2}\rho U^2 A \quad \text{or} \quad U = \sqrt{\left(\frac{W}{C_{\text{L}}\frac{1}{2}\rho A}\right)}. \tag{A.61}$$

Therefore, from eqns (A.60) and (A.61), power

$$P = W\frac{C_{\text{D}}}{C_{\text{L}}}\sqrt{\left(\frac{W}{C_{\text{L}}\frac{1}{2}\rho A}\right)} = \left(\frac{W^3}{\frac{1}{2}\rho A}\right)^{\frac{1}{2}}\frac{C_{\text{D}}}{C_{\text{L}}^{\frac{3}{2}}}. \tag{A.62}$$

For minimum power, $C_{\text{D}}/C_{\text{L}}^{\frac{3}{2}}$ must be a minimum, and since from eqn (A.53) $C_{\text{D}} = C_{\text{Do}}+kC_{\text{L}}^2$ then $C_{\text{Do}}+kC_{\text{L}}^2/C_{\text{L}}^{\frac{3}{2}}$ must also be a minimum. Differentiating and equating to zero gives

whence
$$C_{\text{L}}^{\frac{3}{2}}(2kC_{\text{L}})-\tfrac{3}{2}(C_{\text{Do}}+kC_{\text{L}}^2)C_{\text{L}}^{\frac{1}{2}} = 0,$$

$$kC_{\text{L}}^2 = 3C_{\text{Do}}. \tag{A.63}$$

Putting C_{Lmp} as the lift coefficient for minimum power, then from eqn (A.63),

$$C_{\text{Lmp}} = \sqrt{\left(\frac{3C_{\text{Do}}}{k}\right)}. \tag{A.64}$$

From eqn (A.56), $C_{\text{Lmd}} = \sqrt{(C_{\text{Do}}/k)}$, where C_{Lmd} is the lift coefficient for minimum drag. It then follows that

$$C_{\text{Lmp}} = \sqrt{3}C_{\text{Lmd}}. \tag{A.65}$$

By definition, $C_{\text{Lmp}} = W/\frac{1}{2}\rho U_{\text{mp}}^2 A$ and $C_{\text{Lmd}} = \dfrac{W}{\frac{1}{2}\rho U_{\text{md}}^2 A}$, whence,

and
$$\frac{U_{\text{mp}}^2}{U_{\text{md}}^2} = \frac{C_{\text{Lmd}}}{C_{\text{Lmp}}} = \frac{1}{\sqrt{(3)}}$$

$$\frac{U_{\text{mp}}}{U_{\text{md}}} = \frac{1}{(3)^{\frac{1}{4}}} = 0\cdot76,$$

as text (p. 75).

Appendix 12: Heat conduction through hollow cylinders and spheres (Ch. 7)

Consider a cylinder of length l and internal and external radii r_{i}, r_{o} respectively, and imagine an annulus of thickness dr with a radius r (Fig. A.4). The surface of this annulus will be $A = 2\pi r l$. If the temperature gradient across this annulus is $d\theta/dr$, the rate of heat flow q_{h} across the area A is given by text eqn (7.1):

$$q_{\text{h}} = -\lambda A\frac{d\theta}{dr} = -\lambda 2\pi r l\frac{d\theta}{dr} \quad \text{or} \quad q_{\text{h}}\frac{dr}{r} = -2\pi\lambda l\,d\theta. \tag{A.66}$$

If the temperature of the outer surface is θ_o and that of the inner surface $\theta_i (\theta_o > \theta_i)$, eqn (A.66) can be integrated:

$$\int_{r_o}^{r_1} q_h \frac{dr}{r} = -2\pi l \int_{\theta_o}^{\theta_1} \lambda \, d\theta. \quad \text{(A.67)}$$

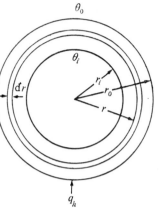

Assuming that q_h and λ are constant, eqn (A.67) gives:

$$q_h = \frac{2\pi\lambda l(\theta_o-\theta_i)}{\ln(r_o/r_i)},$$

as text eqn (7.7).

Using the same procedure for a hollow sphere, the surface area of the annulus will

FIG. A.4

be $4\pi r^2$. If the internal and external radii are r_i, r_o and the corresponding temperatures θ_i, θ_o respectively, eqn (A.66) becomes

$$q_h = -4\pi r^2 \lambda \frac{d\theta}{dr},$$

whence

$$q_h \frac{dr}{r^2} = -4\pi\lambda \, d\theta.$$

On integrating, this gives

$$q_h \int_{r_o}^{r_i} \frac{dr}{r^2} = -4\pi\lambda \int_{\theta_o}^{\theta_i} d\theta,$$

whence

$$q_h = 4\lambda(\theta_o-\theta_i) \Big/ \left(\frac{1}{r_i}-\frac{1}{r_o}\right)$$

as text eqn (7.9).

Appendix 13: Non-steady heat flow (Ch. 7)

Imagine two parallel planes, A and B, each of unit area lying at right angles to the direction of heat flow (A to B), and let the distance between the planes be dx (Fig. A.5). If the temperature gradient at $A = d\theta_1/dx$, then according to text eqn (7.2), the flux of heat ϕ_{hA} through A will be

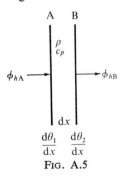

FIG. A.5

$$\phi_{hA} = -\lambda \frac{d\theta_1}{dx}.$$

Suppose that as a result of heat flowing through it, the temperature of the medium rises at the rate of $d\theta/dt$. The plates being of unit area, the volume of medium enclosed is dx. If the density of the medium is ρ and its specific heat c_p, the rate at which heat is taken up by the medium $= \rho c_p \, dx(d\theta/dt)$. Because of this, less heat flows through plane B; if the temperature gradient at B is $d\theta_2/dx$, then the flow of heat through B is

$$\phi_{hB} = -\lambda \frac{d\theta_2}{dx}.$$

The difference in heat flow between A and B must equal the rate at which heat was taken up by the medium, i.e.

$$\rho c_p \, dx \frac{d\theta}{dt} = \phi_{hA} - \phi_{hB} = -\lambda\left(\frac{d\theta_1}{dx} - \frac{d\theta_2}{dx}\right),$$

whence

$$\frac{d\theta}{dt} = -\left(\frac{\lambda}{\rho c_p}\right)\left(\frac{d\theta_1}{dx} - \frac{d\theta_2}{dx}\right) \Big/ dx \quad \text{or} \quad \frac{d\theta}{dt} = \frac{\lambda}{\rho c_p} \cdot \frac{d^2\theta}{dx^2},$$

as text eqn (7.11).

Appendix 14: Freezing of a pond (Ch. 7)

Let $\rho_I =$ density of ice at $0\,°C$ and λ its thermal conductivity. At any instant t s from the start of freezing, let the thickness of the ice be l. If after a further interval of time δt, the thickness has increased by δl, then for *unit surface area*, the volume of new ice formed is δl, its mass is $\rho_I \, \delta l$, and the quantity of heat flowing through the layer above $= L_t \rho_I \, \delta l$, where L_t is the latent heat of fusion of ice at $0\,°C$.

The rate of heat flow through unit area of ice $= L_t \rho_I (\delta l / \delta t)$. If the lower surface of the ice is at $0\,°C$ and the temperature above the ice remains steady at $-\theta$, the temperature gradient is θ/l. From text eqn (7.4)

$$L_t \rho_I \frac{\delta l}{\delta t} = \lambda \frac{\theta}{l}$$

and in the limit ($\delta l, \delta t \to 0$), $dl/dt = \lambda\theta/L_t\rho_I l$. Rearranging and integrating gives

$$\int_0^l l \, dl = \int_0^t \frac{\lambda\theta}{L_t\rho_I} \, dt, \quad \text{whence} \quad l^2 = \frac{2\lambda\theta t}{L_t\rho_I},$$

as text eqn (7.14).

Appendix 15: Alternative expressions for mass transfer and humidity (Ch. 8)

STP $=$ standard temperature of $0\,°C$ (273 K) and pressure of 1 atmosphere $= 1 \cdot 013$ bar.

Molar concentration

For a concentration c (kg m^{-3}) of a substance with a molecular weight M (kg), the molar concentration $c_M = c/M$ (mol m^{-3}). If the mass diffusion rate is q_M mol s^{-1}, text eqn (8.1) can be written

$$q_M = \frac{dc_M}{dt} = -DA\frac{d(c/M)}{dx} = -\frac{DA}{M}\frac{dc}{dx}. \qquad \text{(A.68)}$$

Ideal gas law

For most practical purposes, gas behaviour in biological systems can be assumed to follow the universal gas law:

$$pV = NRT, \qquad \text{(A.69)}$$

where p is the gas pressure, V is the volume occupied by N moles of gas, and T is the absolute temperature (in K $= °C+273$). If p is expressed in bars and V in m^3, we know that at $T = 273$ (0 °C) and atmospheric pressure ($p = 1{\cdot}013$ bar $= 1{\cdot}013\times10^5$ N m^{-2}), the volume occupied by 1 mole of gas (V/N) is the same for all gases $= 2{\cdot}24\times10^{-2}$ m^3, whence R, the gas constant, $= 8{\cdot}31$ J mol^{-1} K^{-1}. A check on the units will confirm that if R is to be expressed in joules, so must be pV; since J $=$ Nm, then if V is in m^3, p must be in N m^{-2} (1 N m^{-2} $= 10^{-5}$ bar).

Data on gases are frequently quoted in terms of their density (ρ kg m^{-3}) which is identical with their mass concentration c. If m is the mass of gas (kg) in a volume V(m^3), $\rho = m/V$ (kg m^{-3}), and if M is the molecular weight of the gas (kg), the number of moles $N = m/M$. Substituting in eqn (A.69) with p *expressed in bars* gives

$$p\frac{m}{\rho} = \frac{m}{M}RT,$$

whence

$$p = \frac{\rho RT}{M} \quad \text{or} \quad \rho = \frac{pM}{RT}. \qquad \text{(A.70)}$$

In other words, at a given temperature T the density of a gas is proportional to the product of its pressure and molecular weight. This applies to both total and partial pressures.

Partial pressures

According to Dalton's law, each gas constituent of a mixture exerts a partial pressure independently of the other constituents and the sum of the partial pressures of each constituent equals the total pressure of the mixture. The partial pressure of a gas is most simply determined from its volume percentage, e.g. if O_2 comprises 21 per cent of the volume of air, its partial pressure will be 21 per cent of the total air pressure. This follows from the fact that each constituent gas occupies the same volume as the mixture so that for a given mass, its density will be directly proportional to its volume percentage and hence, from eqn (A.70) its partial pressure also.

If we substitute for $c \ (= \rho)$ from eqn (A.70) into text eqn (8.1) with

$$q_c = \frac{dm}{dt},$$

we obtain

$$q_c = -DA\frac{M}{R}\cdot\frac{d(p/T)}{dx}. \tag{A.71}$$

Thus, instead of a concentration (density) gradient we have a pressure or partial pressure gradient. In an isothermal system (T constant), the gradient becomes simply dp/dx.

Air humidity

Air humidity can be expressed in various ways:

Absolute humidity ρ_v is the mass of water vapour per unit *volume* of moist air (kg m^{-3}). As noted above this is identical with the density of the water vapour at the prevailing temperature and with the concentration c of water vapour in air.

Specific humidity q is the mass of water vapour per unit *mass* of moist air (dimensionless). Since the mass of moist air (kg) = volume (m^3) × density of moist air ρ_a (kg m^{-3}), then

$$\rho_v = q\rho_a. \tag{A.72}$$

Mixing ratio χ is the mass of water vapour per unit mass of *dry* air (dimensionless). In most environments, the mass of water vapour is negligible compared with the mass of moist air so that in practice, χ and q are usually treated as identical.

Vapour pressure e is the partial pressure of water vapour in air. This can be related to the absolute humidity (ρ_v) by putting $p = e$ into eqn (A.70). With e expressed in mb ($\approx 10^{-3}$ atm), $R = 8\cdot31$ J mol^{-1} K^{-1} and M for water $= M_w = 18 \times 10^{-3}$ kg, we have

$$\rho_v(= c) = \frac{2\cdot17\times10^{-6}}{T}\cdot e \quad \text{(kg m}^{-3}\text{).} \tag{A.73}$$

e can also be related to the specific humidity q by eqn (A.70). We see that

$$\frac{\text{density of water vapour in air } (\rho_v)}{\text{density of moist air } (\rho_a)} = \frac{M_w e}{M_a p},$$

where $M_w =$ molecular weight of water $= 18 \times 10^{-3}$ kg, M_a is the molecular weight of moist air $\approx 29 \times 10^{-3}$ kg, and p is the total air pressure. Since from eqn (A.72), $q = \rho_v/\rho_a$, then

$$q \approx \frac{18\times10^{-3}e}{29\times10^{-3}p} \approx 0\cdot622\frac{e}{p}. \tag{A.74}$$

The ratio e/p is independent of the units if the same units are used. From text eqn (8.35), the psychrometric constant $\gamma = pc_p/0.622L$; therefore,

$$q = c_p e/\gamma L. \tag{A.75}$$

Vapour pressure deficit (vpd) is the difference between the vapour pressure of the air e and the saturation vapour pressure at the same temperature, e_s

$$vpd = (e_s - e).$$

Relative humidity h is the ratio of vapour pressure of the air e to the saturation vapour pressure e_s at the same temperature $= e/e_s$ expressed as a fraction. Very often, e/e_s is expressed as a percentage $= h \times 100$ per cent.

Gases in solution

The *Bunsen coefficient* of absorption α is the volume of gas reduced to STP dissolved by unit volume of solvent at the given temperature in equilibrium with the gas in the gaseous phase at a standard pressure (or partial pressure) of 1 atm ($= 1.013$ bar). If the partial pressure of the gas in the gaseous phase is p (bar), the dissolved volume V_0 is

$$V_0 = \frac{\alpha p}{1.013} \quad \text{and} \quad \alpha = \frac{1.013}{p} V_0. \tag{A.76}$$

The above assumes that the gas obeys Henry's law, namely, that the solubility of the gas is proportional to its pressure. This is valid for most practical purposes.

Suppose that air at a pressure of 1.013 bar, and containing by volume 79% N_2, 20.96% O_2, and 0.04% CO_2, is shaken up with water at 20 °C. At this temperature, the saturation vapour pressure, i.e. the partial pressure of water vapour in saturated air, is 0.023 bar (Fig. 9.10). The total pressure of the N_2, O_2, and CO_2 will then be $(1.013 - 0.023)$ bar $= 0.990$ bar and the partial pressures (see above) will be N_2: (0.79×0.99) bar $= 0.782$ bar; O_2: 0.208 bar; CO_2: 0.0004 bar (total 0.99 bar).

At 20 °C (Table 8.2) α for $N_2 = 0.0164$, for $O_2 = 0.0309$, for $CO_2 = 0.878$. From eqn (A.76) the amounts dissolved (m^3 per m^3 solution) at STP will be:

$$N_2 = \frac{0.0164 \times 0.782}{1.013} = 0.01266; \quad \text{for} \quad O_2 = 0.00635; \; CO_2 = 0.00034$$

This gives a total volume of dissolved gas $= 0.01935$ m^3 per m^3. Expressed as a percentage of this volume we have 65.4% N_2, 32.8% O_2, 1.76% CO_2.

The *Ostwald solubility coefficient* β is the volume of gas, measured at the temperature and pressure at which the gas dissolves, taken up per unit volume of liquid. If β (m^3) is the volume of dissolved gas per m^3 liquid at an absolute temperature T and partial pressure p (bars) in the gaseous phase, then assuming that it obeys the gas law, it can be converted to α at STP by

$$\alpha = \beta \frac{273}{T} \cdot \frac{p}{1 \cdot 013}. \tag{A.77}$$

From eqn (A.76)

$$\beta = \frac{\alpha T}{273} \cdot \frac{1 \cdot 013}{p} \tag{A.78}$$

Appendix 16: Non-steady state diffusion of mass (Ch. 8)

Imagine two parallel planes A and B, *each of unit area* and distance δx apart, at right-angles to the direction of diffusion x (Fig. A.6). If the concentration gradient at $A = dc_1/dx$ then according to text eqn (8.1), the mass of substance diffusing across A in unit time (the flux ϕ_c) is given by

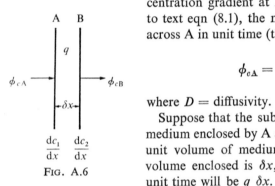

$$\phi_{cA} = -D \frac{dc_1}{dx},$$

where D = diffusivity.

FIG. A.6

Suppose that the substance is consumed in the medium enclosed by A and B at a rate q (mass per unit volume of medium in unit time). Since the volume enclosed is δx, the amount consumed in unit time will be $q\,\delta x$. If as a result of this, the gradient at B is dc_2/dx, then the mass of substance diffusing across B in unit time is given by

$$\phi_{cB} = -D \frac{dc_2}{dx}.$$

It follows that

$$q\,\delta x = -D\left(\frac{dc_1}{dx} - \frac{dc_2}{dx}\right),$$

whence

$$q = -D\left(\frac{dc_1}{dx} - \frac{dc_2}{dx}\right) \bigg/ \delta x.$$

In the limit ($\delta x \to 0$), this becomes

$$q = D \frac{d^2c}{dx^2} \quad \text{or} \quad \frac{q}{D} = \frac{d^2c}{dx^2}.$$

Instead of q, we can express the rate of consumption as a rate of change in concentration in the medium with time, i.e. $q = dc/dt$, whence

$$\frac{dc}{dt} = D \frac{d^2c}{dx^2},$$

as text eqn (8.18).

A solution to eqn (8.18) obtained by integrating twice is

$$c = \frac{1}{2}\frac{q}{D}x^2 - c_1 x + c_0,$$
(A.79)

where c_1 and c_0 are integration constants. When $x = 0$, $c = c_0$.

Applying the above to the diffusion of O_2 through a tissue, with q as the rate of O_2 consumption, c as the concentration of O_2 at a depth x from the surface, and c_0 as the concentration of O_2 at the surface ($x = 0$). If the thickness of the tissue is such that at the centre ($x = l/2$), the supply of O_2 is just limiting respiration so that there is no further change in its concentration, then $dc/dx = 0$. Differentiating eqn (A.79) and putting $dc/dx = 0$ when $x = l/2$, gives

$$\frac{q}{D}\cdot\frac{l}{2} - c_1 = 0 \quad \text{and} \quad c_1 = \frac{ql}{2D}.$$
(A.80)

Substituting for c_1 in eqn (A.79) then gives

$$c = c_0 - \frac{qx}{2D}(l - x),$$

as text eqn (8.19).

Appendix 17: Radiation geometry (Ch. 9)

The account given below is taken largely from Monteith (1973).

In Fig. A.7, A_h represents the area of shadow cast by a body with surface area A on a horizontal surface by a beam of radiation inclined at an angle β to the horizontal.

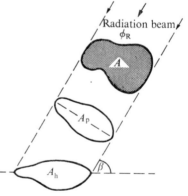

If A_p is the area of A projected normal to the beam, then

$$A_p = A_h \sin \beta.$$

If ϕ_R is the radiation flux, the intercepted radiation is given by

$$A_p\phi_R = \phi_R(A_h \sin \beta) = A_h E,$$

where $E = \phi_R \sin \beta =$ irradiance of surface occupied by the shadow, i.e. intercepted radiation = area of shadow on horizontal plane $\times$ irradiance of that plane.

Fig. A.7

The total radiation intercepted $\bar{E}$ is given by

$$\bar{E} = \left(\frac{A_h}{A}\right)E.$$

The ratio A_h/A is known as the *shape factor* and can be calculated geometrically or measured directly from the shadow. For a sphere $E = 0{\cdot}25\phi_R$.

Appendix 18: Solarimeters and net radiometers (Ch. 9)

The most commonly used solarimeters (pyranometers) consist of a thermopile, blackened to ensure maximum absorption of solar radiation, and protected by a glass dome (transparent only to solar radiation) (see Fig. A.8). The absorption of radiation by one side of the thermopile (i.e. one set of thermojunctions) induces a temperature difference proportional to the amount of solar energy absorbed; this induces a corresponding thermoelectric current which is measured by a sensitive galvanometer or can be recorded.

The net radiometer consists essentially of two identical solarimeters placed back to back and connected in opposition so that the net thermoelectric

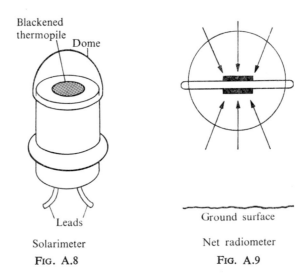

Solarimeter

FIG. A.8

Net radiometer

FIG. A.9

output corresponds to the difference in radiant energy absorbed by the two thermopiles. With the radiometer mounted above the ground (Fig. A.9) so that the upper thermopile receives all incoming radiation and the lower, all outgoing radiation from the surface, the difference corresponds to the *net radiation* R_N. In order to ensure that all radiation is received (i.e. both short-wave and long-wave) the glass dome is replaced by thin polythene kept inflated by nitrogen gas from a cylinder.

Appendix 19: EPER formula (Ch. 9)

The wet and dry bulb eqn (8.35) with $e = e_a$ and $e_s = e_{sw}$, is

$$e = e_s - \gamma(\theta - \theta_w), \tag{A.81}$$

where e is the vapour pressure of the air at the dry bulb temperature θ,

e_s the saturation vapour pressure at the wet bulb temperature θ_w, and γ the psychrometer constant.

If the wet and dry bulb temperatures are measured at two heights and their differences expressed as $\Delta\theta$ and $\Delta\theta_w$, and the same applied to e and e_s, then

$$\Delta e = \Delta e_s - \gamma(\Delta\theta - \Delta\theta_w). \tag{A.82}$$

Rearranging gives

$$\frac{1}{1+\gamma(\Delta\theta/\Delta e)} = \frac{\Delta e}{\Delta e_s + \gamma\,\Delta\theta_w} = 1 - \frac{\gamma\,\Delta\theta}{\Delta\theta_w\{(\Delta e_s/\Delta\theta_w)+\gamma\}}.$$

Substituting $s = \Delta e_s/\Delta\theta_w$ gives

$$\frac{1}{1+\gamma(\Delta\theta/\Delta e)} = 1 - \left(\frac{\gamma}{s+\gamma}\right)\left(\frac{\Delta\theta}{\Delta\theta_w}\right). \tag{A.83}$$

From eqn (A.75) and text eqn (9.28),

$$\frac{H}{LE} = \gamma\,\frac{\Delta\theta}{\Delta e} \quad \text{and} \quad LE = \frac{R_N - G}{1+\gamma(\Delta\theta/\Delta e)}.$$

Substituting from eqn (A.83) gives

$$LE = (R_N - G)\left\{1 - \left(\frac{\gamma}{s+\gamma}\right)\left(\frac{\Delta\theta}{\Delta\theta_w}\right)\right\},$$

as text eqn (9.29).

Appendix 20: Evaporation from free water surfaces and leaves (Ch. 9)

1. *Free water surfaces*

We begin with text eqns (9.30b) and (9.30d),

$$H = \frac{\rho c_p(\theta_1 - \theta_2)}{r_{ah}} \quad \text{and} \quad LE = \frac{\rho c_p(e_1 - e_2)}{\gamma r_{aw}}.$$

If measurements are made at height z and at the surface, $\theta_1 = \theta_0$, $\theta_2 = \theta_z$, $e_1 = e_{s0}$, the saturation vapour pressure at the surface temperature θ_0, and $e_2 = e_z$, whence

$$H = \frac{\rho c_p(\theta_0 - \theta_z)}{r_{ah}} \tag{A.84a}$$

and

$$LE = \frac{\rho c_p(e_{s0} - e_z)}{\gamma r_{aw}}. \tag{A.84b}$$

From eqn (A.84b),

$$e_{s0} = \frac{\gamma r_{aw} LE}{\rho c_p} + e_z. \tag{A.85}$$

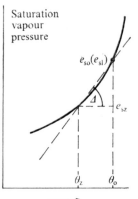

Saturation vapour pressure

$e_{so}(e_{sl})$

Δ

e_{sz}

θ_z θ_o

Temperature $\theta °C$

FIG. A.10

Over a small part of the temperature–saturation vapour pressure curve (Fig. A.10), the slope Δ is given by

$$\Delta = \frac{e_{s0}-e_{sz}}{\theta_0-\theta_z},$$

whence

$$\theta_0-\theta_z = \frac{e_{s0}-e_{sz}}{\Delta}, \tag{A.86}$$

where e_{sz} is the saturation vapour pressure of the air at the height z. Substituting for $\theta_0-\theta_z$ in eq (A.84a) gives

$$H = \frac{\rho c_p(e_{s0}-e_{sz})}{\Delta r_{ah}}. \tag{A.87}$$

From text eqn (9.10), $LE = (R_N-G)-H$. Substituting for H from eqn (A.87) gives

$$LE = (R_N-G)-\frac{\rho c_p(e_{s0}-e_{sz})}{\Delta r_{ah}}.$$

Substituting for e_{s0} from eqn (A.85) gives

$$LE = (R_N-G)-\frac{\rho c_p}{\Delta r_{ah}}\left(\frac{\gamma r_{aw}LE}{\rho c_p}+e_z-e_{sz}\right) = (R_N-G)-\frac{\gamma r_{aw}LE+\rho c_p(e_z-e_{sz})}{\Delta r_{ah}}.$$

Rearranging gives

$$LE\left(1+\frac{\gamma}{\Delta}\frac{r_{aw}}{r_{ah}}\right) = (R_N-G)+\frac{\rho c_p}{\Delta r_{ah}}(e_{sz}-e_z)$$

or

$$LE = \frac{\Delta(R_N-G)+(\rho c_p/r_{ah})(e_{sz}-e_z)}{\Delta+\gamma r_{aw}/r_{ah}},$$

as text eqn (9.31). If $r_{aw} = r_{ah} = r_a$, then

$$LE = \frac{\Delta(R_N-G)+(\rho c_p/r_a)(e_{sz}-e_z)}{\Delta+\gamma},$$

as text eqn (9.32).

2. *From leaves*

Eqn (9.30b) for heat transfer between a leaf surface (θ_0) and air at a height $z(\theta_z)$ gives

$$H = \frac{\rho c_p(\theta_0-\theta_z)}{r_{ah}}. \tag{A.88}$$

Eqn (9.35) for evaporation between the mesophyll (e_{sl}) and air at a height $z(e_z)$ gives

$$LE = \frac{\rho c_p(e_{sl}-e_z)}{\gamma(r_s+r_{aw})}, \tag{A.89}$$

whence

$$e_{sl} = \frac{\gamma(r_s + r_{aw})}{\rho c_p} LE + e_z. \tag{A.90}$$

As shown in Fig. A.10, the slope of the saturation vapour pressure–temperature curve,

$$\Delta = \frac{e_{sl} - e_{sz}}{\theta_0 - \theta_z}. \tag{A.91}$$

Substituting for $(\theta_0 - \theta_z)$ from eqn (A.91) into eqn (A.88) gives

$$H = \frac{\rho c_p (e_{sl} - e_{sz})}{\Delta r_{ah}}. \tag{A.92}$$

Substituting for e_{sl} from eqn (A.90) into eqn (A.92) gives

$$H = \frac{\rho c_p}{\Delta r_{ah}} \left\{ \frac{\gamma(r_s + r_{aw})LE}{\rho c_p} + e_z - e_{sz} \right\}. \tag{A.93}$$

From text eqn (9.10), $LE = (R_N - G) - H$, whence

$$LE = (R_N - G) - \frac{\gamma LE (r_s + r_{aw})}{\Delta r_{ah}} + \frac{\rho c_p}{\Delta r_{ah}} (e_{sz} - e_z).$$

Rearranging gives

$$LE = \frac{\Delta(R_N - G) + (\rho c_p / r_{ah})(e_{sz} - e_z)}{\Delta + \gamma(r_{aw} + r_s)/r_{ah}}.$$

If $r_{aw} = r_{ah} = r_a$, then

$$LE = \frac{\Delta(R_N - G) + (\rho c_p / r_a)(e_{sz} - e_z)}{\Delta + \gamma(1 + r_s/r_a)},$$

as text eqn (9.36).

Appendix 21: Air diffusion resistance and wind velocity (Ch. 9)

Text eqn (9.30a) represents the transport of momentum between two heights (z_1, z_2) at which the mean horizontal velocities are respectively u_1 and u_2. If $z_1 = d + z_0$, i.e. $z_1 - d = z_0$, then from text eqn (9.15), $u_1 = 0$, and eqn (9.30a) becomes

$$\tau = \frac{\rho u_2}{r_{am}} \quad \text{or} \quad r_{am} = \frac{\rho u_2}{\tau}.$$

Assuming that the flux of momentum is constant with height,

$$\frac{\tau}{\rho} = \frac{\tau_0}{\rho} = u_*^2,$$

whence $r_{am} = u_2/u_*^2$. Putting $u_1 = 0$, $u_2 = u_z$, $z_1 = d+z_0$, and $z_2 = z$ into text eqn (9.25) and squaring gives

$$u_*^2 = (u_z)^2 \bigg/ \left\{ 2 \cdot 5 \ln\left(\frac{z-d}{z_0}\right) \right\}^2,$$

whence

$$r_{am} = \left\{ 2 \cdot 5 \ln\left(\frac{z-d}{z_0}\right) \right\}^2 \bigg/ u_z,$$

as text eqn (9.38).

References

ALEXANDER, R. M. (1968). *Animal mechanics*, Sidgwick and Jackson, London.

ALLEN, R. D. and FRANCIS, D. W. (1965). Cytoplasmic contraction and the distribution of water in the *Amoeba*. In *The state and movement of water in living organisms*, (ed. G. E. Fogg), p. 259. S. E. B. Symposium No. 19. Cambridge University Press.

BAINBRIDGE, R. (1963). Caudal fin and body movement in the propulsion of some fish. *J. exp. Biol.* **40**, 23.

BENNETT, C. O. and MYERS, J. E. (1962). *Momentum, mass, and heat transport.* McGraw-Hill, New York.

BERNAL, J. D. (1965). The structure of water and its biological implications. In *The state and movement of water in living organisms*, (ed. G. E. Fogg) p. 17. S.E.B. Symposium No. 19. Cambridge University Press.

BIRD, R. B., STEWART, W. E., and LIGHTFOOT, E. N. (1960). *Transport phenomena.* John Wiley, New York.

BIRKEBAK, R. C. (1966). Heat transfer in biological systems. *Int. Rev. gen. exp. Zoology*, **2**, 269.

BRIGGS, G. E. and ROBERTSON, R. N. (1948). Diffusion and absorption in discs of plant tissue. *New Phyt.* **47**, 265.

BROWN, H. T. and ESCOMBE, F. (1900). Static diffusion of gases and liquids in relation to the assimilation of carbon and its translocation in plants. *Phil. Trans. R. Soc. B.* **193**, 223.

BROWN, R. H. J. (1963). The flight of birds. *Biol. Rev.* **38**, 460.

BURTON, A. C. (1972). *Physiology and biophysics of the circulation.* (2nd edn.) Year Book Medical Publishers, Chicago.

CABORN, J. M. (1957). Shelterbelts and microclimate. *Forestry Commission Bulletin* No. 29. H.M.S.O. London.

CANNY, M. J. (1973). *Phloem translocation.* Cambridge University Press.

CARMAN, P. C. (1956). *Flow of gases through porous media.* Butterworth's Sci. Pub.

CARSLAW, H. S. and JAEGER, J. C. (1969). *Conduction of heat in solids.* (2nd edn.) Clarendon Press, Oxford.

CHADWICK, L. E. (1953). The motion of wings; aerodynamics and flight metabolism. In *Insect physiology* (ed. K. D. Roehder) pp. 577 and 615. Chapman and Hall, London.

CHARM, S. E. and KURLAND, G. S. (1974). *Blood flow and microcirculation.* John Wiley, New York.

——, PALTIEL, B., and KURLAND, G. S. (1968). Heat transfer coefficients in blood flow. *Biorheology*, **5**, 133.

CHURCH, N. S. (1960). Heat loss and the body temperatures of flying insects. II. Heat conduction within the body and its loss by radiation and convection. *J. exp. Biol.*, **37**, 186.

CLARKSON, D. T., RICHARDS, A. W., and SANDERSON, J. (1971). The tertiary endodermis in barley roots; fine structure in relation to radial transport of ions and water. *Planta*, **96**, 292.

COKELET, G. R. (1972). The rheology of human blood. In *Biomechanics* (ed. Y. C. Fung, N. Perrone, and M. Anliker), p. 63. Prentice Hall.

CRONSHAW, J. and ANDERSON, R. (1969). Sieve plate pores of *Nicotiana. J. Ultrastruct. Res.*, **27**, 134.

DAINTY, J. (1963). Water relations of plant cells. *Adv. bot. Res.* **1**, 279.

DIXON, H. H. (1914). *Transpiration and the ascent of sap in plants.* Macmillan, New York.

DIXCN, M. (1951). *Manometric methods.* 3rd edn. Cambridge University Press.

DROST-HANSEN, W. (1969). Water near solid interfaces. *Ind. engng. Chem.*, **61**, 10.

DUSHMAN, S. (1962). *Scientific foundations of vacuum technique.* 2nd edn. John Wiley, New York.

EDE, A. J. (1967). *An introduction to heat transfer.* Pergamon Press, London.

EISNER, F. (1930). *3rd Int. Cong. appl Mech. Stockholm* (*see* Rohsenow and Choi 1961).

EXLEY, R. R., BUTTERFIELD, B. G., and MEYLAN, B. A. (1974). Preparation of wood specimens for the scanning electron microscope. *J. Microscopy*, **101**, 21.

FAHRAEUS, R. and LINDQUIST, T. (1931). The viscosity of blood in narrow capillary tubes. *Am. J. Physiol.*, **96**, 562.

FARMER, J. B. (1918). On the quantitative differences in the water conductivity of the wood in trees and shrubs. *Proc. R. Soc. B*, **90**, 218.

FENN, W. O. (1927). The oxygen consumption of frog nerve during stimulation. *J. gen. Physiol.*, **10**, 767.

FISHENDEN, M. and SAUNDERS, O. A. (1961). *An introduction to heat transfer.* Clarendon Press, Oxford.

FRASER, A. I. (1963). Wind tunnel studies on the forces acting on the crowns of small trees. In *Forestry Commission Report on Forest Research*, 1962. H.M.S.O. London.

FREY-WYSSLING, A. (1952). *Deformation and flow in biological systems.* North Holland, Amsterdam.

GARDNER, W. R. (1960). Soil water relations in arid and semi-arid conditions. *Arid zone Res.*, **15**, 37.

GATES, D. M. (1962). *Energy exchange in the biosphere.* Harper and Rowe.

——, KEEGAN, H. J., SCHLETER, J. C., and WEIDNER, V. E. (1965). Spectral properties of plants. *Appl. Optics*, **4**, 11.

GEIGER, R. (1965). *The climate near the ground.* (English translation). Harvard University Press.

GERRARD, R. W. (1931). Oxygen diffusion into cells. *Biol. Bull.*, **60**, 245.

GLOYNE, R. W. (1954). Some effects of shelterbelts on local and micro-climate. *Forestry*, **27**, 85.

GOLDSTEIN, S. (1938). *Modern developments in fluid dynamics*, vol. II. Clarendon Press, Oxford.

GRAY, J. (1968). *Animal locomotion.* Wiedenfeld and Nicolson, London.

GREENWALT, C. H. (1962). Dimensional relationships for flying animals. *Smithsonian misc. Collns*, **144**, No. 2.

HANCOCK, G. J. (1953). Self-propulsion of microscopic organisms through liquids. *Proc. R. Soc. A*, **217**, 96.

HEINE, R. W. (1970). Estimation of conductivity and conducting area of poplar stems using a radioactive tracer. *Ann. Bot.*, **34**, 1019.

—— (1971). Hydraulic conductivity in trees. *J. exp. Bot.*, **22**, 503.

HERRINGTON, L. P. (1969). On temperature and heat flow in tree stems. *Bulletin* 73, Yale University School of Forestry.

HERTL, H. (1966). *Structure, form and movement*. (English translation). Rheinhold, New York.

HILLEL, D. (1971). *Soil and water; physical principles and processes*. Academic Press, New York.

HOUSE, C. R. (1974). *Water transport in cells and tissues*. Edward Arnold, London.

HYASHI, T. (1961). How cells move. *Sci. Am.*, **205**, 184.

JAHN, T. L. and VOTTA, J. J. (1972). Locomotion of protozoa. *Ann. Rev. fluid Mech.* **4**, 93.

JAKOB, M. (1949). *Heat transfer*, vol. 1. John Wiley, New York.

JARMAN, P. D. (1974). The diffusion of carbon dioxide and water vapour through stomata. *J. exp. Bot.*, **25**, 927.

KAMIYA, N. (1942). Physical aspects of protoplasmic streaming. In *The structure of protoplasm*, (ed. W. Seifriz), p. 199. *Symp. Am. Soc. Plant Physiologists*, Iowa State College Press.

KATCHALSKY, A. and CURRAN, P. F. (1967). *Non-equilibrium thermodynamics in biophysics*. Harvard University Press.

KLOTZ, I. M. (1970). Water: its fitness as a molecular environment. In *Membranes and ion transport*, (ed. E. E. Bittar), p. 93, vol I. Wiley Interscience, New York.

KRAMER, M. O. (1965). Hydrodynamics of the dolphin. *Adv. Hydroscience*, **2**, 111.

KRAMER, P. J. (1969). *Plant and soil water relationships*. McGraw-Hill, New York.

KUTATELADZE, S. S. (1963). *Fundamentals of heat transfer*. (English translation). Edward Arnold, London.

LANG, T. G. and PRYOR, K. (1966). Hydrodynamic performance of porpoises (*Stenella attenuata*). *Science*, **152**, 531.

LEONARD, E. R. (1939). Studies on tropical fruits VI. A preliminary consideration of the solubility of gases in relation to respiration. *Ann. Bot.* **3**, 825.

LEYTON, L. (1971). Problems and techniques in measuring transpiration from trees. In *Physiology of tree crops* (ed. L. C. Luckwill and C. V. Cutting), p. 101. Academic Press, New York.

LIGHTHILL, M. J. (1960). Note on the swimming of slender fish. *J. Fluid Mech.*, **9**, 305.

LOWRY, W. P. (1969). *Weather and life; an introduction to bio-meteorology*. Academic Press, New York.

McILROY, I. C. (1971). An instrument for continuous recording of natural evaporation. *Agric. Meteorol.*, **9**, 93.

McINTOSH, D. H. and THOM, A. S. (1969). *Essentials of meteorology*. Wykeham, London.

MARSLAND, D. A. (1942). Protoplasmic streaming in relation to gel structure in the cytoplasm. In *The structure of protoplasm*, (ed. W. Seifriz), p. 127. *Symp. Am. Soc. Plant Physiologists*. Iowa State College Press.

MAYHEAD, G. J. (1973). Some drag coefficients for British forest trees derived wind tunnel studies. *Agric. Meteorol.*, **12**, 123.

MEIDNER, H. and MANSFIELD, T. A. (1968). *Physiology of stomata*. McGraw-Hill, New York.

MEYLAN, B. A. and BUTTERFIELD, B. G. (1971). *3-dimensional structure of wood*. Chapman and Hall, London.

MILLER, P. L. (1966). The supply of oxygen to the active flight muscles of some large beetles. *J. exp. Biol.*, **45**, 285.

MILLINGTON, R. J. and QUIRK, J. P. (1960). Transport in porous media. *7th Int. Cong. Soil Sci. Madison, Wisconsin*, p. 97, vol. I(3).

MONTEITH, J. L. (1965). Evaporation and the environment. In *The state and*

226 References

movement of water in living organisms, (ed. G. E. Fogg). *S.E.B. Symp.* No. 19, p.205. Cambridge University Press.

—— (1972). *Instruments for micrometeorology.* Blackwell Scientific Publishers, Oxford.

—— (1973). *Principles of environmental physics.* Edward Arnold, London.

NACHTIGALL, W. (1964). Die Schwimmechanik der Wasserinsekten. *Ergebn. Biol.*, **27**, 39.

—— (1966). Die Kinematik der Schlagflugelbewegungen von Dipteren. *Z. vergl. Physiol.*, **52**, 155.

—— and BILO, D. (1965). Die Strömungsmechanik des Dytiscus-rumpfes. *Z. vergl. Physiol.*, **50**, 371.

NAGELI, W. (1946). Weitere Untersuchungen über die Windverhältnisse im Bereich von Windschutzanlagen. *Mitteil. Schweiz. Anst. Forstl. Versuchsw.*, **24**, 659.

—— (1953). Untersuchungen über die Windverhältnisse im Bereich von Schilfrohr-wänden. *Mitteil. Schweiz. Anst. Forstl. Versuchsw.*, **29**, 213.

OTIS, A. B., FENN, W. O., and RAHN, H. (1950). Mechanics of breathing in man. *J. appl. Physiol.*, **2**, 592.

PAPPENHEIMER, J. R. (1953). Passage of molecules through capillary walls. *Physiol. Rev.*, **33**, 387.

PARLANGE, J-Y. and WAGGONER, P. E. (1970). Stomatal dimensions and resistance to diffusion. *Plant Physiol.*, **46**, 337.

—— and HEICHEL, G. H. (1971). Boundary layer resistance and temperature distribution on still and flapping leaves. I. Theory and laboratory experiments. *Plant Physiol.*, **48**, 437.

PASQUILL, F. (1950). The aerodynamic drag of grassland. *Proc. R. Soc. A*, **202**, 143.

PENMAN, H. L. (1948). Natural evaporation from open water, bare soil and grass. *Proc. R. Soc. A*, **193**, 120.

—— (1956). Vegetation and hydrology. Tech. Comm. No. 53, Commonwealth Bureau of Soils, Harpenden, Commonwealth Agricultural Bureaux.

—— and SCHOFIELD, R. K. (1951). Some physical aspects of assimilation and transpiration. In *Carbon dioxide fixation and photosynthesis*, *S.E.B. Symp.* No. 5, p. 115. Cambridge University Press.

PENNYCUICK, C. J. (1972). *Animal flight.* Inst. Biol. Studies in Biol. No. 33, Edward Arnold, London.

PETTY, J. A. (1970). Permeability and structure of the wood of Sitka spruce. *Proc. R. Soc. B*, **175**, 149.

—— (1972). The aspiration of bordered pits in conifer wood. *Proc. R. Soc. B*, **181**, 395.

PRANDTL, L. (1952). *Fluid dynamics.* Blackie, London and Glasgow.

—— and TIETJENS, O. G. (1934). *Applied hydro- and aerodynamics.* (English trans-lation 1st edn). McGraw-Hill, New York.

PRINGLE, J. W. S. (1957). *Insect flight.* Cambridge University Press.

PROTHERO, J. and BURTON, A. C. (1961). The physics of blood flow in capillaries. I. The nature of the motion. *Biophys. J.* **1**, 565.

—— (1962). II. The capillary resistance to flow. *Biophys. J.*, **2**, 199.

RAPP, G. M. (1970). Convective mass transfer and the coefficient of evaporative heat loss from human skin. In *Physiological and behavioral temperature regulation*, (ed. J. D. Hardy, A. P. Gagge, and J. A. J. Stolwijk), p. 55. C. C. Thomas, Springfield, Illinois.

RASPET, A. (1961). Biophysics of bird flight. Ann. Rept. Board Regents, Smith-sonian Inst. Year ending 1960, p. 405.

REID, R. C. and SHERWOOD, T. K. (1966). *The properties of gases and liquids.* (2nd edn). McGraw-Hill, New York.

REYNOLDS, O. (1883). An experimental investigation of the circumstances which determine whether the motion of water shall be direct or sinuous, and of the law of resistance in parallel channels. *Trans. R. Soc.*, **174**, 935.

ROHSENOW, W. M. and CHOI, H. Y. (1961). *Heat, mass, and momentum transfer.* Prentice-Hall.

RUSSELL, E. W. (1973). *Soil conditions and plant growth.* (10th edn). Longmans, London.

SAYRE, J. D. (1926). Physiology of stomata of *Rumex patientia. Ohio J. Sci.*, **26**, 233.

SCHOLANDER, P. F. (1955). Climatic adaptation in homeotherms. *Evolution*, **9**, 15.

—— (1957). The wonderful net. *Sci. Am.* **196**, 97.

—— (1958). The rise of sap in lianas. In *Physiology of forest trees* (ed. K. V. Thimann), p. 3. Ronald Press, New York.

——, HAMMEL, H. T., BRADSTREET, E. D., and HEMMINGTON, E. A. (1965). Sap pressure in vascular plants. *Science*, **148**, 339.

SCHULTZ, R. D. and ASUNMAA, S. K. (1970). Ordered water and the ultrastructure of the cellular plasma membrane. *Recent Prog. Surf. Sci.*, **3**, 291.

SEIFRIZ, W. (1942). *The structure of protoplasm. Symp. Am. Soc. Plant Physiologists*, Iowa State College Press.

SELLERS, W. D. (1965). *Physical climatology.* University of Chicago Press.

ŠESTÁK, Z., CATSKY, J., and JARVIS, P. G. (1971). *Plant photosynthetic production; manual of methods.* Dr. W. Junk, The Hague.

SIAU, J. F. (1971). *Flow in wood.* Syracuse University Press.

SLATYER, R. O. (1967). *Plant–water relationships.* Academic Press, New York.

SPANNER, D. C. (1962). A note on the velocity and energy requirement of translocation. *Ann. Bot.*, **26**, 511.

—— (1970). The electro-osmotic theory of phloem transport in the light of recent measurements on *Heracleum* phloem. *J. exp. Bot.*, **21**, 325.

STEHBENS, W. E. (1960). Turbulence of blood flow in the vascular system of man. In *Flow properties of blood* (ed. A. L. Copley and G. Stainsby). p. 137. Pergamon Press, Oxford.

STOKES, R. H. and MILLS, R. (1965). *Viscosity of electrolytes and related properties, International Encyclopedia of Chemistry and Chemical Physics*, vol. 3. Pergamon Press, Oxford.

SUTTON, O. G. (1953). *Micrometeorology.* McGraw-Hill, New York.

SZEICZ, G. and LONG, I. F. (1969). Surface resistance of crop canopies. *Water Resources Res.*, **5**, 622.

——, ENDRODI, G., and TAJCHMAN, S. (1969). Aerodynamic and surface factors in evaporation. *Water Resources Res.*, **5**, 380.

TANI, N. (1963). The wind over the cultivated field. *Bull. natn. Inst. agric. Sci., Tokyo, Ser. A*, **10**, 74.

TANTON, T. W. and CROWDY, S. H. (1972). Water pathways in higher plants. II Water pathways in roots. *J. exp. Bot.*, **23**, 600.

TAYLOR, G. I. (1951). Analysis of the swimming of microscopic organisms. *Proc. R. Soc. A*, **209**, 447.

—— (1952a). Action of waving cylindrical tails in propelling microscopic organisms. *Proc. R. Soc. A*, **211**, 225.

—— (1952b). Analysis of the swimming of long and narrow animals. *Proc. R. Soc. A*, **214**, 158.

THOM, A. S. (1968). The exchange of momentum, mass, and heat between an artificial leaf and the airflow in a wind tunnel. *Quart. J. R. Met. Soc.* **94,** 44.

TIBBALS, E. C., CARR, E. K., GATES, D. M. and KREITH, F. (1964). Radiation and convection in conifers. *Am. J. Bot.,* **51,** 529.

TREGEAR, R. T. (1965). Hair density, windspeed and heat loss. *J. appl. Physiol.,* **20,** 796.

—— (1966). *Physical functions of skin.* Academic Press.

TYREE, M. T. (1971). The steady state thermodynamics of translocation in plants. In *Trees, structure, and function.* (M. H. Zimmermann and C. L. Brown), p. 281. Springer-Verlag, Berlin.

——, DAINTY, J. and BENIS, M. (1973). The water relations of hemlock (*Tsuga canadensis*). I Some equilibrium water relations as measured by the pressure-bomb technique. *Can. J. Bot.,* **51,** 1471.

VAND, V. (1948). Viscosity of solutions and suspensions. I. Theory. *J. phys. Colloid Chem.,* **52,** 277.

VOGEL, S. (1967). Flight in *Drosophila.* III. Aerodynamic characteristics of fly wings and wing models. *J. exp. Biol.,* **46,** 431.

—— (1970). Convective cooling at low air speeds and the shapes of broad leaves. *J. exp. Bot.,* **21,** 91.

—— and BRETZ, W. L. (1972). Interfacial organisms: passive ventilation in the velocity gradients near surfaces. *Science,* **175,** 210.

WALSHE, D. E. and FRASER, A. I. (1963). Wind tunnel tests on a model forest. *National Physical Laboratory Aero Report,* No. 1078.

WEATHERLEY, P. E. and JOHNSON, R. P. C. (1968). The form and function of sieve tubes: a problem in reconciliation. *Int. Rev. Cytol.,* **24,** 149.

WEBB, E. K. (1965). Aerial microclimate. *Meteorol. Monogr.,* **6,** 27.

WEIHS, D. (1973a). Hydromechanics of fish schooling. *Nature,* **241,** 290.

—— (1973b). The mechanism of rapid starting of slender fish. *Biorheology,* **10,** 343.

WEIS-FOGH, T. (1972). Energetics of hovering flight in hummingbirds and in *Drosophila. J. exp. Biol.,* **56,** 79.

—— (1973). Quick estimates of flight fitness in hovering animals including novel mechanisms for lift production. *J. exp. Biol.,* **59,** 169.

WHITMORE, R. L. (1968). *Rheology of circulation.* Pergamon Press, Oxford.

van WIJK, W. R. and DE VRIES, D. A. (1963). Periodic temperature variations in a homogeneous soil. In *Physics of plant environment.* (ed. W. R. van Wijk). p. 102. North Holland, Amsterdam.

WOOD, C. J. (1973). The flight of the albatross (a computer simulation). *Ibis,* **115,** 244.

W.M.O. (World Meteorological Organisation) (1964). Windbreaks and shelterbelts. *Technical Note* 59, Geneva.

—— (1968). Agricultural meteorology. 2 vols. *Proceedings of seminar, Melbourne,* 1966. Published by Bureau of Meteorology, Australia.

YAKUWA, R. (1945). Über die Bodentemperaturen in dem verschiedenen Bodenarten in Hokkaido. *Geophys. Mag.* (*Tokyo*), **14,** 1.

ZIMMERMANN, M. H. (1971). Transport in the xylem. In *Trees, structure, and function* (ed. M. H. Zimmermann and C. L. Brown), p. 169. Springer-Verlag, Berlin.

The following books, though not quoted in the text, have proved valuable as additional sources of data:

DUNCAN, W. J., THOM, A. S., and Young, A. D. (1970). *Mechanics of fluids* (2nd ed.). Edward Arnold, London.

MASSEY, B. S. (1970). *Mechanics of fluids* (2nd ed.). Van Nostrand, New York.

HORNE, R. A. (ed.) (1972). *Water and aqueous solutions: structure, thermodynamics and transport processes.* Wiley Interscience, New York.

LIGHTFOOT, E. N. (1974). *Transport phenomena and living systems.* John Wiley, New York.

YOST, W. (1960). *Diffusion in solids, liquids and gases* (3rd ed.). Academic Press, New York.

Subject Index

Reference Data in Tables and Figures